Development Strategies for Barak Valley (Assam)

Development Strategies for Barak Valley (Assam)

Edited by

J B Bhattacharjee

AKANSHA PUBLISHING HOUSE
NEW DELHI-110002 (INDIA)

AKANSHA PUBLISHING HOUSE
4649-B/21, Ansari Road
Darya Ganj, New Delhi - 110 002
Ph.: 23263193 / 9811582579 / 09435116718
Fax: 011-23263193
Email: ektabooks@yahoo.com
www.akanshapublishinghouse.com

Development Strategies for Barak Valley (Assam)

First Published 2009
ISBN 978-81-8370-176-1

PRINTED IN INDIA

Published by Akansha Publishing House, New Delhi and Printed at Aryan Enterprises, Delhi.

Contents

Preface

The Barak Valley in Assam, consisting of three districts, viz. Cachar, Hailakandi and Karimganj, is one of the economically more backward parts of the country. It is a small geographical area of about seven thousand square kilometers with a population of almost thirty lakhs according to the census of 2001.This landlocked and truncated portion of the erstwhile Surma Valley is virtually sandwiched between the hills on three sides and the international boundary with Bangladesh on the fourth. Transferred to Assam from the Dacca Division of Bengal in 1874 (when Assam was constituted into a province) but separated from the Assam proper by the hill districts, this predominantly Bengali-speaking area is geographically, historically and culturally a distinct region in the State of Assam. Its economy was shattered by the partition of India in 1947 when the major part of the valley (i.e. Sylhet) went to Pakistan and its road, rail and waterways to the rest of the country were snapped away.

The book emerges from a brainstorming session on 'Development Strategies for Barak Valley : Constraints and Prospects' organized by the Institute of Northeast India Studies in collaboration with the Department of Economics, Assam University, Silchar, on 29th April 2008 at the sprawling and vibrating campus of the University. The brainstorming was planned in four parts, namely, inauguration, presentation, interaction and valediction. The brief inaugural session early in the morning, in which the keynote address formed the main part, was followed by the presentations to which the entire forenoon was devoted. As many as fourteen presenters made their submissions in that session covering various aspects of the problem of economic development of the Barak Valley and the strategies that would be appropriate for the area. The afternoon was devoted to interactions in which a good number of participants opened themselves up in frankly sharing their personal views on the problems of development and the possible resolutions with or without reference to the presentations made in the forenoon. The valedictory session in the evening engaged itself at a resume of the daylong academic discourse. The interactions and deliberations in various sessions were voice-recorded and the rapporteur could compare her own copious notes of the proceedings with the recorded version for preparation of a postscript on the discourse. We expect the book to be useful to all those who are involved in the development processes in the country in general and in Northeast India in particular, because we sincerely believe that no problem can be resolved without understanding its root causes. The discourse revealed that the intelligentsia in the Barak Valley is convinced that the economic backwardness of the valley, despite its enormous potential of growth, is mainly due to the indifference and apathy of the government.

Institute of Northeast India Studies deeply appreciates the support received from various quarters in organizing the brainstorming session and in publishing the proceedings in a book form. The initiative for the brainstorming was by our two colleagues in the Assam University – Dr Projit Kumar Palit and Dr Niranjan Roy. The Department of Economics, Assam University, proposed to hold the brainstorming session and the University gave them full support. The teachers and research scholars in the Department of Economics were fully involved in the programme and they mobilized participation of the colleagues in other departments and affiliated colleges and the representatives of the local intelligentsia. Dr Niranjan Roy indeed excelled in his role as the organizing secretary. The leadership to the whole endeavour came from the Vice-Chancellor, Professor Tapodhir Bhattacharjee, who inspired the programme by his leadership from the front with determination to ensure that the Assam University plays its role in the socio-economic, cultural and intellectual advancement of the area under charge of the University, true to the commitment under Statute of the University. The Vice-Chancellor chaired the inaugural as well the valedictory session and thus kept himself available from the beginning to the end of the brainstorming and repeatedly reminded his colleagues in his presidential remarks about the role the University should play in the development of the region. Professor G D Sharma, Pro-Vice Chancellor, delivered the valedictory address, while Professor Gopalji Misra, Dean, School of Social Sciences, also addressed the concluding session. Both the addresses were directly on the central theme of the brainstorming and enlivened the discourse by highlighting the key issues. Professor Keya Sengupta, a former Head of the Department of Economics and a former Dean of the School of Social Sciences, Assam University, who chaired the presentations, made the brainstorming absorbingly interesting by her timely interventions and moderations and by her own contributions to the discourse. The high academic standard of the brainstorming session was possible due to the presenters and the participants. They braved the sensitivity of the subject and offered their balanced and objective views in highest academic spirit. The local press and the channels joined the effort by extensively covering the event. Finally, Shri M P Misra, the enterprising proprietor of the Akansha Publishing House, New Delhi, took the responsibility of publishing and brought out the book in good time.

Institute of Northeast India Studies
Kolkata - 700129
September 11, 2008

(J B Bhattacharjee)

Keynote

—J B Bhattacharjee

It is more or less an accepted fact that the districts (namely, Cachar, Hailakandi and Karimganj) in the Barak Valley segment of Assam are amongst the backward districts in India, both in terms of geographical location and the state-of-art in economic development. It is also historically true that when India emerged as a republic after two hundred years of the British colonial rule, she was among the backward or poor countries – impoverished and pauperized by colonial extractions and exploitation, but in sixty years since independence the country has certainly made rapid strides in various fields of development and today it is undoubtedly one of the fast developing nations. The claim that India is already an economic super-power may be contested, but her spectacular progress in agricultural and industrial growth and her achievement in science and technology are beyond questioning. Similarly, the Northeast is considered as a backward region of the country and for many years we have been complaining against the indifference of the central government towards the development of the region. But today, the things are different. The development of the Northeast is now getting special attention of the Government of India. A number of special programmes have been launched and, if sincerely implemented, this region is also expected to forge ahead in the development processes. The impact of the changed policy is already becoming apparent on the surface in some parts of the region in the form of improved highways, roads and bridges, broad-gauge railway, revamped airports, impressive buildings, new institutions, offices, agency-houses, factories, urban growth, etc, and the symptoms of freshness in the overall life style of the people, particularly in the urban agglomerations. Therefore, the question is : If the rest of India can progress and if the rest of the Northeast, including the rest of Assam, can also reasonably move forward, why not the Barak Valley ? It can no longer be denied that the Northeast is now getting special attention and liberal funding for development. The Northeast Package, Non-Lapsable Pool, revamping of NEC, and finally, a separate Union Ministry for the Development of the North-Eastern Region (DONER), are examples. It is generally held that Northeast should be able to make best use of these special funding to develop itself and be at par with the rest of the country, that the special funding of this nature cannot continue eternally, and that if the region, or any part thereof, fails to take advantage of these special dispensations in a time-frame, it shall be difficult for those backward areas of the Northeast to develop themselves in future. The question is: Whether Barak Valley is getting appropriate share of these special dispensations for development ?

The constraints of development for the Barak Valley are indeed many. Sandwiched between the hills and the international boundary, the density of population in this landlocked tiny part of the erstwhile Surma Valley (of the British period and Srihatta of the early times) is among the highest in the country. It is a distinct culture zone, different from the rest of Assam and the neighbouring north-eastern states, and the continuity is only with the adjoining old Srihatta portion of Tripura and Bangladesh. Its continuing ills began with the partition of India in 1947 when it was separated from the rest of the valley to which it geographically, culturally and historically belonged. The partition also snapped away its shortest and natural routes to the rest of India (East Pakistan/ Bangladesh coming in between). With the transfer of the larger portion of the valley to Pakistan, this smaller portion lost its political voice as well. On the other hand, population increased beyond sustainable limit as an immediate impact of the partition. This population growth told heavily on the local resources, settlements grew abruptly, forests were depleted, and the agricultural land shrunk in size. The low-yielding outdated agricultural techniques probed to be painfully inadequate to match the growing demand for foodstuff. The valley became rice-short and dependant on supply from outside. There was no remarkable effort to scientifically modernize the agricultural productivity. The package programme dawned and perished as a paper-tiger. In fact, the holdings in the valley are so small that the agrarian reforms and modernization of agriculture seem to be difficult without a region-specific scientific approach and a sincere effort on the part of the political authority.

The flood has been a perennial problem for the people of the valley, but the flood control measures are believed to be only half-hearted and temporary. The Barak Dam, which is heard of since 1950s, remains a distant dream. Whether any project in this country has taken so much time to fruit as the Barak Dam can be an interesting subject of academic enquiry. The Barak Dam was conceptualized as a multipurpose project to control flood, to provide irrigation facility and to generate hydro-electricity. It is now given to understand that the delay is due to objections raised by a neighbouring state and a neighbouring country. These objections deserved timely intervention, technical scrutiny by the experts and appropriate alternations in the site and the design so that the project does not harm the neighbouring areas in any way. The development of an area should not be allowed to adversely affect its neighbours. On the other hand, a dam or a barrage in the river Barak is a must for containing floods and for the economic development of Barak Valley. The technology today is capable of taking precautions against all possible adversities, provided there is no dearth of sincerity on the part of the government. It is generally believed that the matter has been complicated by interfering with the original concept of the dam and by delaying the execution. Like the Barak Dam, the extension of the broad-gauge railhead (Lumding to Silchar) is also taking more time than anywhere else. In this case, the failure of the government to provide adequate security to the construction agencies is understood to be the cause of delay.

The logical approach to development of Barak Valley could be industrialization, but no sincere effort is so far noticed in that direction. The techno-economic survey reports in early 1950s highlighted the abundant locally available raw-materials for sugar and pulp industries, but the public sector sugar mill is a case of infant mortality and the paper mill has been struggling to achieve the targets. The gas-crackers experiments are too tiny for the optimum utilization of the huge natural gas deposits in the valley as it is understood from the ONGC sources. It is indeed a pity that the natural resources of the area had to remain unexplored for such a long time. The tea industry is almost exhausted, primarily due to loss of natural fertility of soil, lack of initiative on the part of the management in replenishing the plants and scientific fertilizing and cultivation. In fact, the local economy had never a chance to gain much from tea which was in no stage of its development different from a classical colonial capitalist enterprise. In this so-called industry, only land was local but all other factors of production – labour, capital and organization – were from outside. On the other hand, it brought tremendous pressure on the local economy. The tea-planters in the British time took more land than what could be used for the purpose for which the grants were made and the estate-holders retained the excess land till parts of those could be recovered in early 1970s under ceiling-surplus regulations. It is indeed sad that even after independence there were attempts to glorify this colonial initiative by campaigning that tea earns foreign exchange, and the question - what portion of that foreign exchange came back to the region? - was lost in emotionally surcharged discourses. Nonetheless, the development of the tea industry to help it recover its old glory deserves special schemes urgently, and at the same time, the possibility of a more returnable alternative for the utilization of the unutilized land and the surplus labour in the moribund tea estates should be an important agenda in formulating the development strategies.

The rest of India may rejoice the 'glorious sixty years since independence', but in Barak Valley it is generally held an era of stagnation and deterioration in many fronts. Some of the very good roads of the British time, including the old Silchar-Hailakandi Road that passes through the Assam University Campus (which was known as Cachar Trunk Road) and the Silchar-Sylhet Road via Karimganj (which was called Sylhet Trunk Road), are indeed in a deplorable condition due to indifference and apathy of the government towards maintenance of the roads. The Silchar-Karimganj sector looks somewhat tolerable only because of its integration in the national highways, but the condition of the rest of the roads, including those within the towns and municipalities, have been allowed to dwindle over the years. The towns have grown and expanded haphazardly without any planning and supervision. The conditions of the hospitals and educational institutions are equally awesome. Most of the schools and colleges for general education were established by private initiative. The infra-structure is too poor to match the requirements of quality education and other demands of the 21st century. The government has been failing to look after even the primary schools. Some of the prestigious institutions of the valley have lost their old pride for the same reason. The problem of unemployment, including the plight of the educated youth, is mounting up over the years without being addressed seriously. The local sources of energy generation have not been exploited and the supply that is obtained from outside is notoriously short of requirement even for domestic consumption. The power situation is the best example to explain why the valley failed to attract corporate investments for industrial growth despite tremendous possibilities.

The development strategies have to be considered in the context of the problems and the resource-based prospects. In any case, industrialization shall have to be taken up seriously if the valley has to survive the global economic challenges of the 21st century. The tea estates have to be protected against continual sickening, and the agriculture must be modernized. Two other prospective areas are (i) commerce and (ii) human resource. The geographical location of the valley has endowed it with a centre-stage both in the regional and the cross-country trade and commerce across the Northeast. The fact that five of the seven states in the Northeast meet in Silchar (within few kilometers from the city) might explain that centre-stage paradigm, while Bangladesh is intimately bordered and Myanmar (or the access to Southeast Asia) is also not far. The challenge is to develop appropriate infrastructure to gain out of the proposed 'Look East' policy and the Asian Highway. The other area, namely, the human resource development also becomes important in this specific context. The entrepreneurial training and management education are especially important in the emerging scenario. Tourism development, involving scientific preservation of the heritage sites and training the youth in tourism, hotel and event managements can be another potential area. The universities and other institutions of higher, technical and professional education are expected to play specialized roles in this new dispensation. The time is fast running out for the mounting problem of educated unemployment to be addressed seriously and scientifically.

The next logical question is: Who will do all these ? In sixty years since independence the general impression in the valley is that the state government – though the government of the state – has not been able to probe its sincerity for the development of this part of the state. The alternative proposed by some organizations – as it is known from a section of the press – is a sort of a regional body to look after the economic development, autonomous and empowered to generate resources, to receive central assistance and to monitor the development projects. It is understood that the Barak Upatyaka Bahu-Bhasik Samanway Samiti has already submitted a memorandum to the Prime Minister to this effect and that an organization under the name 'Barak Valley Economic Development Council Demand Committee' has also been formed to pursue this and many other long standing demands of the people of the valley for economic development. The question is: What should be the constitutional or legal status and the method of instituting such a body? Shall the proposed 'economic development council' be a democratic and constitutional authority like the district and regional councils as provided in 5th and 6th Schedules of the Constitution of India and thereby another state-like organization within the state, or, should it be a body of experts, or a mix of democracy, bureaucracy and technocracy ? The answers to these queries are not very clear from what came out in the newspapers. However, we may get those answers in this brainstorming. It may also be recalled that the concept of an autonomous economic development council for Barak Valley is older than the recent initiative. It emerged as an academic proposal when two economists of Assam University – Professors P R Bhattacharjee and P Nayak – argued the rationale of economic autonomy for Barak Valley in a research paper published in the *Journal of the North-East India Council for Social Science Research*, Shillong [P R Bhattacharjee & P Nayak, "Socio-Economic Rationale of a Regional Development Council for the Barak Valley", *JNEICSSR*, vol. 27, no. 1, 2003]. I am glad that the concept of my former colleagues (and published in a Journal, of which I am one of the editors) may be revisited today.

The objective of this brainstorming session is to deliberate on the development strategies for Barak Valley in the context of the constraints of development and on the strength of its potentials of growth. I expect the discourse to incite ideas on specific issues related to the economic development of the valley, and we may discuss those issues dispassionately and comprehensively.

PRESENTATIONS

1

Colonial Continuity in History of Barak Valley

—Sanjib Deb Laskar

The term colony or colonialism has been used for over two centuries, and has been interpreted by many in many ways. In the present time, the word bears a condemnatory inference and is equated to aggression, exploitation and negotiation of human rights. The world had witnessed the rise of colonialism in the 18th century when several European nations undertook to explore the continents, to extend their rule by force, by opening diplomatic and commercial relations, by preaching religion, and by subjugating nations by military aggressions. This colonialism is a multi-faceted concept active in different levels, from within and without; its impact on human life is so deep-rooted that it continues to cut into the bones and marrows of a nation, long after the colonial rule is withdrawn and a nation achieves, what is called, independence. This may be termed as the legacy of colonialism, which leaves its mark in the making of policies and in its implementation in a nation which has achieved political freedom driving out the colonialists by violent revolution or by peaceful accords, or by both simultaneously. In the present world or the post-war world, the spectre of colonialism still subsists and has been disseminated through the mechanisms of governance in the post-colonial nations. Again, under the backdrop of one form of colonialism on a broad spectrum, there emerges a sort of regional colonialism moving downwards towards extreme local level.

I

Coming over to the Barak Valley in South Assam, if we take an overview of the history, the conventional way of demarcation of the historical eras as pre-colonial, colonial and post-colonial tend to lose its significance as one fails to identify different phases to determine the points where colonialism began and where it ended. Located at a peripheral area, the plain riverine valley of Barak (erstwhile Cachar), which experienced social and polity formation at very early stage of history of the Northeast - dating back to the pre-Christian era - had always been under some state powers which were not particularly rooted to this soil. The advent of Arya-Brahminic group or the rise of Tripuri kingdom, or the extension of Samatata, Kamarupa or Harikela from south-east Bengal or the emergence of Shrihattarajjya in the ancient period of history had incorporated the area from the borders of Sylhet to the Barail hill range in the east touching upon the border of Manipur. Although no clear history of the period could yet be reconstructed, it has been reasonably established by the historians of the contemporary times [Bhattacharjee : 1991; Choudhury : 2006] that the valley of Barak (along with that of Surma in Sylhet) had made its stride towards social and polity formations during the period; the forests began to be cleared, the fallow land brought under cultivation, and the process of village formation made considerable progress. However, it is widely accepted that the state-powers mentioned above had extended their rule in this land which, as a matter of fact, was ruled by the external powers, while the people of the soil explored the natural resources on their own and remained content with the feeling that there existed a ruler above to whom an annual tribute has to be paid through his representatives, and this presence of the supreme power itself offered them a sense of security.

There are epigraphic evidences of land-settlement in Sylhet-Cachar region during Samatata, Tripuri and Kamarupi rule in the first millennium of the Christian era. [Mitra: 1880; Choudhury: 2002; Gupta: 1967] This shows that the state power started the process of extension of agriculture in the region. In all these cases, the seat of power had always remained outside the valley, and it is reasonable to believe that the development activities in the valley had in some way or the other a colonial feature. A close reading of the land-grants shows that the valley of Barak (also of Surma) had drawn the attention of the power-centres during the first millennium; its land measured, attempts made to extend wet-rice cultivation by offering grants to the Brahmins. [Choudhury: 2006: 57-67] These Brahmin recipients of the land-grants (by Bhutivarman of Kamarupa) may be looked upon as the early group of settlers in Sylhet-Cachar region who may be credited with the initiation of an agricultural revolution.

The valley had witnessed a power-tussle involving two state powers - Tripura and Koch Bihar - in the mid 16th Century AD, in which the Tripuri lost Cachar plains to the Koch. [Guha : 2006 : 21-29] With Khaspur as capital, the Koch began to rule the 'crown colony', and soldiers from the mainland of Koch Bihar took possession of the land, Koch colonies grew with Rajbanshis or Dehans immigrating there. The first ruler, Kamalnarayan, unlike his military brother Chilarai, represented a different temperament; he engaged himself in social and developmental works like construction of roads, establishment of religious shrines, etc. Nonetheless, under the Koch rule Cachar was a sort of a colony, at least in its early part, when the wealth of the land had been transferred outside, particularly its forest resources. Of course, during the period new villages began to emerge, some sort of community guilds (termed as 'Khel'} began to develop, and artisans from Jayantia and other parts of Sylhet came to settle in Cachar. As many as eight kings ruled in Cachar during the period (of whom, one was assassinated), and in 1745 the Koch rule ended yielding place to the Dimasas.

With formal annexation of Cachar to the Dimasa state (1745) Barak Valley experienced a new system of administration. [Guha ; 2006 : 63] Of course, the Dimasas had already established their foothold in some bordering places prior to it, and by the early 18th Century at the behest of the rulers in Maibong a significant portion of Cachar had been measured, and the details of the lakes, rivers, fallow land were taken note of. We gather it from the reading of an appointment letter delegating the powers of *Uzir* to Maniram Laskar of Barkhola issued by Maharaja Kirtichandranarayan in 1736. [Bhattacharyya : 2003 : 14-16] It appears that long before the Dimasa capital was shifted to Khaspur, the important information, including names of important persons in the valley, were gathered by the Maibong Court. Maniram Laskar was assigned the task of revenue collection, and administrative power was conferred on him with the assurance that whoever would defy him would be dealt with severity. After the shifting of capital, more and more local people were offered important positions in the court. In fact, the participation of the local people in the affairs of the state was a unique feature of the Dimasa administration which differentiated the Dimasa rule from any other sort of colonial administration.

The 82 years of Dimasa rule was a very important period for the economy and the social and cultural life in Cachar, although the Dimasa state had witnessed a gradual rise followed by a rapid decline and ultimate fall. With the help of the indigenous revenue specialists, the Dimasa court had assessed the land revenue, organized land system, offered settlements and created a number of posts like *muktar, khelmuktar, bhandari,* etc., to be filled in by the local people in Barak Valley. Excavation of ponds, construction and maintenance of roads, maintenance of ferry *ghats* and *chowkis*, maintenance of *khels*, exploitation of forest resources including live elephants, ivories, medicinal plants etc, were major activities carried out by the Dimasa court for the economic development of the state. There is a reference to the 'worship of river Barak' at the royal initiative, presumably to protect the land from the ravages of the river, an evidence of the concern for the economy of the land. Considering various aspects of development strategies during the rule, there is reason to believe that the Dimasas had identified themselves to this land with no intention of making it a colony. As the capital had been permanently shifted to the plains, the wealth or the surplus income tended to remain within the valley.

During the last phase of Dimasa rule, Cachar became an arena of extra-territorial power games involving Manipuri princes to which was added the Burmese incursions, followed by the court intrigues. [Guha: 2006: 92-93] The security weakened, and economy declined. A Mughal general Aga Mohmmad Reza occupied a portion of Cachar (1899). Again, a displaced British *subedar* Kalyan Singh attacked Cachar (1803), and the king made a humble prayer to the British to provide him security. Following Krishnachandra's death (1813) and ascension of Govindachandra to the throne, Cachar entered into yet another abject state of crisis. The royal exchequer became empty, revenue collection recorded decrease, heavy tax imposed on the subjects, titles and *miras* sold at random, and all these had an adverse reaction against the king, who had to escape time and again from the capital to avoid attempts on his life. The king, however, received British assistance in 1824 to resist the Burmese attack, and in 1826 entered into an agreement with the British following which the independent kingdom of Cachar virtually lost its sovereignty to the colonialists, agreeing to pay an annual tribute of '10000 Sikka' to them. [Guha : 2006 : 102-04] Following six years of chaos and lawlessness, economic depression paved the way to annexation of the kingdom to the colonial territory of the British to which, however, the majority of the important persons and common people gave their assent. [Guha : 2006]

II

After occupying the kingdom, the British Government declared Cachar as a 'non-regulated province' (according to the Regulation X of 1822) and thereafter, it was included in the Dacca division of the Bengal Presidency (1836). It is important to note that following the British annexation on 14 August 1832, Cachar became a 'province' with Agent to the Governor General, North-East Frontier, Cherrapunji, at the upper level and a Superintendent at the ground level of administration. Captain Thomas Fisher was the Superintendent of Cachar and David Scott, the Agent to the Governor General. This David Scott held three posts simultaneously, namely, Special Commissioner of North-East Rangpur (later on Goalpara district), Commissioner of Assam, and the Agent to the Governor General, North-East Frontier. In the dispatches, Scott was addressed by the Superintendent not as Commissioner of Assam, but as the Agent to the Governor General, North-East Frontier. However, the status of a 'province' was denied to Cachar when it was made a district under the Commissioner of Dacca in Bengal Presidency. From an independent kingdom to a subsidiary state under British paramountcy, thereafter, a non-regulated province, and then a district, - that was the stages of declining status of Cachar under the colonial rule. Cachar was, however, separated from Bengal and tagged with Assam (along with Sylhet) in 1874, and its fate had been subjected to more changes, adjustments, re-adjustments, following the territorial re-distribution of districts (Sylhet and Cachar) in 1905, 1911 and 1947. The transfer and re-distribution of portions of Sylhet-Cachar region (Surma Valley of the British time) were carried out without the consideration of the geographical condition and the linguistic, cultural, and ethnic composition of Cachar and Sylhet. [Bhattacharyya & Laskar: 2000; Gait: 1981: 336-36]

In the matters relating to governance in Cachar under the British, the traditional '*muktari*' system was first replaced by '*tahashildari*' system, which was again replaced by '*muktari*' system between 1832 and 1836, and in 1843 the *tahashildari* system was restored. Such was their desperate move seeking ways to improve revenue to cater to the colonial interest. An observation by the British civilian-historian, E.A.Gait throws some light on the colonial attitude :

> The district officers,...were in almost all cases military officers, transferred from the local regiments to civil employ, and so long their orders were not openly flouted and the revenue was collected with fair punctuality, they left most things in the hands of the subordinates and troubled themselves little with details of the district works. [Gait: 1981 : 339]

III

The introduction of tea plantation in Cachar in 1856 had been a great boon to the colonialists, who, to make most out of this promising industry, gave attention to development of infrastructure, improvement of communication network, and the security arrangements. The industry flourished and the income generated from it was flushed away to the homeland of the British colonialism. The railway line extended to the valley, the roads and communication system improved, waterways connected the district with the outside world, and all these were surely to boost the tea-industry. The growth of urban centres in the region (including Sylhet) owes much to this colonial enterprise. The policies were systematically maintained when the Indian capitalists from the other provinces of British India joined the European planters. The things remained unchanged even after independence of the country, as the practice of making profit and transferring the capital outside without re-investing in the district remained the same. On the other hand, in the post-colonial period also the tea-lobby continued to be a determining factor in politics, administration and the economic life, not only for Cachar but for the whole of Assam. However, it may be a subject of serious academic discourse as to whether the tea sector has been able to shake off the colonial legacy.

IV

The major part of the Sylhet district was separated from the present Barak Valley and handed over to East Pakistan at the termination of the British colonial rule. The left over of the British Surma Valley, i.e. the three districts of the Barak Valley today, remained attached with Assam, a state which represents different traditions, cultures, languages, social systems and ethnic roots, bearing a political agenda not subscribed by the people of Barak Valley. Following partition, Barak Valley had to accommodate a large number of displaced people from across the border, and the valley passed through serious economic crises as the relief and rehabilitation works were inadequate. The conflict of interests between Brahmaputra Valley and Surma Valley (which included Cachar and Sylhet), termed as 'valleyism', had already created a distance between the two in British time, and the changed circumstances after partition put the Barak Valley at a more disadvantageous position, as the people in Barak Valley were weakened numerically and politically by the transfer of Sylhet.

Such being the state of things, the Cachar District Congress Committee was not ready to come within Assam

Pradesh Congress Committee {APCC} [Choudhury: 2007 : 51] and still wanted to cling to the Bengal Pradesh Congress Committee (BPCC), although the geographical connections with Bengal (West Bengal) were already cut off by the partition of India. The congress leaders of Assam, from Brahmaputra Valley, at the immediate pre-partition days, were not in favour of retaining Cachar with Assam, and would not mind had it been handed over to Pakistan along with Sylhet. Under the circumstances, the position of Cachar after independence was in no way better than in the colonial days. The APPC under G.N. Bardoloi in 1946 had already made it clear in the election manifesto that they were much more concerned with the well being of their own community than that of the others The Manifesto stated [Choudhury : 2007 : 51]

> Unless the province of Assam be organized on the basis of Assamese language and culture, the survival of the Assamese nationality and culture will become impossible. The inclusion of Bengali... Sylhet and Cachar (plains portion) and the immigration or importation of lacs of Bengali settlers on wastelands have been threatening to destroy the distinctiveness of Assam.

Only 22 days before the Independence of the country, Ambika Giri Roy Choudhury had issued a statement in *The Assam Tribune* (July 22, 1947) that:

There is a little cause in trying to retain the junior partner of Sylhet - the Cachar plains, at any rate, Hailakandi sub division, in Assam.

This has been reiterated by another supporter of the group in a letter the next day, as:

> There seems to be no justification in retaining a few thanas of Sylhet district and Hailakandi sub division of Cachar district within the province of Assam. (*The Assam Tribune*, July 23, 1947)

The position of Cachar (Barak Valley) in Free India becomes clearer when one reads the Governor of Assam's speech at the first Assembly session held at Shillong on November 5, 1947, prepared and approved by the Bordoloi Ministry:

> The natives of Assam are now masters of their own house. They have a Government which is both responsible and responsive to them. They can take what steps are necessary for the encouragement and propagation of Assamese language and culture and the language and customs of the tribal people who are their follow citizens and who also must have a share in the formation of such policies. The Bengalee has no longer the power, even if he had the will to impose any thing on the people of the hills and valleys which constitute Assam. The basis of such feeling against him as exist is fear but now there is little or no cause of fear. I would therefore appeal to you to exert all influence to give the stranger in our midst a fair deal, provided, of course, he in his turn deals loyally with us.

On May 8, 1948, an instruction was issued by the Government of Assam, the text of which is given below:

> In view of the emergency created by the huge influx of refugees into the province from Pakistan territories and in order to preserve peace, tranquility and social equilibrium in the towns and villages, Government reiterate their policy that settlement of lands should in no circumstances be made with persons who are not indigenous to the province. The non-indigenous inhabitants of province should include, for the purpose of land settlement during the present emergency, persons who are non-Assamese settlers in Assam though they already have lands and houses of their own and have made Assam their home to all intents and purposes.

Again, a definition of 'indigenous person' was given in the *Assam Gazette* dated 6 September, 1950, wherein the existence of Bengali language or linguistic group has been kept vague. The lines read thus:

> Indigenous persons of Assam and speaking Assamese language or any tribal dialect of Assam, or in the case of Cachar, the language of the region.

This willful omission of the Bengali language speaks a volume of the intention of the provincial power-centre against the peripheral marginalized region. This attitude is reflected not only in the matters relating to land settlement, but also in the field of education too. The government decided to deprive the students of the facility of scholarship on this ground. The circular reads thus:

> Government of Assam have decided that they would not henceforth be responsible for the payment of any scholarship in any institution to those students who do not have domicile in Assam, as constituted after the transfer of a portion of Sylhet to East Bengal. (Circular No. 27868-933, dated 10.09.48., DPI, Assam)

The policies of the provincial government after independence failed to reflect the hopes and aspirations of the peripheral areas like Cachar or Barak Valley and also of the different districts of undivided Assam under British India. The seeds of future discontent were sown by the policies of the government which were to manifest in the form of movements in Assam asserting regional, ethnic, linguistic identity resulting in balkanization of the province and showing the possibility of still more segregation to enable different sections determine their own fate removing the spectre of colonialism.

REFERENCES

Bhattacharjee, J.B, (1991), *Social and Polity Formations in Pre-Colonial North-East India: The Barak Valley Experience,* Vikas Pub. House, New Delhi.

Bhattacharyya Jayanta Bhusan & Sanjib Deb Laskar (2000) "Barak Upatyaka: Bahaman Itihaser Dhara" (in Bengali), *Samayik Prasanga,* 13 Nov. 2000.

Bhattacharyya, Padmanath (2003), *Hedamba Rajyer Dandabidhi Ebong Hedamba Rajyer Rindanbidhi* (in Bengali) ed. Amalendu Bhattacharyya, Barak Upatyaka Banga Sahitya O Sanskriti Sammelan, Silchar (First published 1911).

Choudhury, Achyuta Charan (2002), *Shrihatter Itibritta* (in Bengali), vol. I, New Edition, Utsha Prakashan, Dhaka (First published 1910).

Choudhury, Sujit (2006), *Shrihatta Cacharar Prachin Itihas*, (Bengali), Dinkal Press, Silchar.

Choudhury, Sujit (2007), *Barak Upatyakar Samaj O Rajniti* (in Bengali), Jugashakti Prakashan, Karimganj.

Gait, E.A (1981), *A History of Assam*, LBS Pub., Guwahati (First published, 1905).

Guha, Upendra Chandra (2006), *Cacharer Itibritta* (in Bengali), New edition edited by Amalendu Bhattacharyya & Sanjib Deb Laskar, Parul Prakashan, Agartala (First published, 1912).

Gupta, Kamalakanta (1967), *Copper-plates of Sylhet,* Sylhet

Mitra, Rajendralal (1880), "Copper-Plate Inscriptions from Sylhet", *Proceeding of the Asiatic Society of Bengal*, No VIII, August 1880.

2

Constraints of Development of Barak Valley

—Girban Biswas

Development has been a catch word in modern democracy and a favourite topic to generate storm over the cup of tea in the middle class society. It generally means the raising of the standard of life of the people by proper harnessing of natural resources of the concerned regions. Similarly, the stereotype of understanding the constraints of development would mean the non-availability of proper infrastructure and adequate resources, absence of skilled manpower, and so on. It may be a departure from this traditional paradigm of constraints, but even at the risk of such aberrations, in case of the Barak Valley in Assam it is necessary to delve into the history, geography, society and economy for an understanding of the constraints of development of the region.

We do not intend to go to the remote antiquity to narrate the history of the Barak Valley, for it may suffice to say that it has a long history of being ruled by different rulers, along with Sylhet or in parts, since ancient times. On the eve of the British occupation (which also occurred in parts), the Cachar plains formed part of the small Dimasa state which had not yet seen the true development of its productive forces and production relations as well as the optimum exploitation of its natural resources. The economy and society of the area was linked with the neighbouring Sylhet *Sircar* (district) of the Bangala Subah of the Mughals as much as it was connected by the geography. The British took it over in 1832 and made it a district under the Dacca Division of Bengal (where Sylhet was already another district). The Cachar and Sylhet districts (thereafter, together called 'Surma Valley Division') of the Dacca Division of Bengal were transferred to the new province of Assam which was created as a Chief Commissioner's Province in 1874. In fact, the British used Cachar-Sylhet as a sacrificial goat for the viability of Assam as a frontier province and both the districts were made to move like a shuttlecock between Bengal and Assam a number of times till 1947. The final assault was in 1947 when Sylhet district (except the area that now forms the Karimganj district in Barak Valley) was handed over to Pakistan following the partition of India. This partition shattered its economy, took away the natural routes of communication, imposed a cultural and linguistic isolation, and enfeebled the political voice of the truncated valley which had played a glorious role in India's struggle for independence. Cachar had to bear the major burnt of the partition and huge displacement of people. By its separation from Bengal and making a part of Assam, and finally handing over the Sylhet part of the valley to Pakistan, Cachar became seriously handicapped. One may imagine the consequence of the separation of a major limb from the main body and the compulsion of living a desolate and isolated life. This division of the main body of the valley itself turned out to be an insurmountable constraint for the development of Barak Valley.

Geographically, the Barak Valley is a landlocked region, crisscrossed by a large number of rivers and lakes (locally called, *beel* or *haor*) and surrounded by hills and mountains, with its only natural outlet through Sylhet. The creation of

Pakistan (now Bangladesh) and the separation of Sylhet from India blocked the natural outlet not only for the people of Barak Valley but also its rivers, roads and railways. The disturbance to the natural drainage system is resulting in the recurrence of flood havoc for the valley almost every year. The forced union of the Bengali majority Barak Valley with the Assamese majority province of Assam politically weakened the people of the valley. The leadership of Barak Valley can hardly play any significant role in the political life of the province due to its insignificant numerical strength. On the other hand, there are internal contradictions within the local society in the valley which are prone to be magnified in the events of political mobilization. The composite ethnic character of the Barak Valley, including the religious polarization between the Bengali Hindus and the Bengali Muslims, proved to be detrimental to the formation of a sound lobby or pressure group either for demanding political autonomy or resisting the prolonged discrimination by the political leadership of the Brahmaputra Valley.

The geographical isolation is indeed a major constraint for economic development of Barak Valley. The communication system is highly deplorable by any standard. The moribund rail link established by the British in the nineteenth century and a fragile road for surface transport across the hills are life lines for the valley for the movement of the people and the supply of essential commodities. On the other hand, the landlocked valley with insufficient production of essential commodities depends almost entirely on the supply from outside. The situation is a stimulus for the greedy trading class to charge exorbitant price for essential commodities. The cost of living in this part of Assam is said to be highest in the country and this is attributed to the communication bottleneck.

Economically, the infrastructure in power, communication, agriculture and industry is dismally poor. The want of state patronage through irrigation, agricultural restructuring, agrarian reform, and overall appropriate development initiative prevented the economy from a take off. One pertinent factor for poor economic progress is the absence of proactive initiative to control and to prevent the recurrence of frequent floods in the valley. The mainstay of economy of the valley is agriculture, but the agricultural productivity depends entirely on fair weather. The government virtually did nothing either to provide the irrigation facility to the cultivators or to protect their paddy fields from the flood havoc. The incidences of the cultivators growing paddy in the fields to be washed away by devastating floods before harvesting are very common experiences in the valley. This cruel destruction is compelling the traditional cultivators to look for alternative means of livelihood. The need to tame the river Barak was felt long ago, and immediately after independence the people in the valley raised the demand for a dam in Barak. This 'Barak Dam' is heard of since 1950s. However, considering the renewed demand of the people of the valley, the Government of India a few years back decided to construct a dam at Tipaimukh. The foundation stone of this Tipaimukh Project has also been laid, but there is hardly any visible sign of work being undertaken in the site. It is, therefore, doubtful whether this project will ever be completed. Consequently, the prospect of agricultural development in the Barak Valley also seems to be gloomy.

Whereas the indifferent attitude of the state government in Assam towards needs and aspirations of the people is generally held as a major constraint for the development of Barak Valley, a civil society initiative for wresting development from the post-colonial Indian state should also be held equally guilty for its conspicuous absence. The thirty lakh population in the Barak Valley consists of four major linguistic and religious groups, namely, Bengali Hindu, Bengali Muslim, Manipuri, and tea garden and ex-tea garden workers. The Manipuris include Meitei and Bishnupriya. There are also some small tribal groups. Inspite of living together for centuries a cohesive unity of mind and understanding has not yet developed between these groups within the valley. This might be the reason why there is no democratic movement for realization of the popular demands like extension of the broad gauge railway, road networking, industrialization, flood control and agricultural development. The people of the valley must understand that unless a strong civil society emerges a landlocked area like the Barak Valley is bound to remain 'locked' in all aspects of development.

REFERENCES

Bhattacharjee, J B (1977), *Cachar Under British Rule in North East India*, Radiant Publishers, New Delhi.

Bhattacharjee, J B (ed) (1989), *Sequence of Development in North East India*, Omsons Publishing House, New Delhi.

Choudhury, Achyuta Charan (2002), *Shrihatter Itibritta* (in Bengali), vol. I, New Edition, Utsha Prakashan, Dhaka (First published 1910).

Choudhury, Sujit (2006), *Shrihatta Cacharar Prachin Itihas*, (Bengali), Dinkal Press, Silchar.

Choudhury, Sujit (2007), *Barak Upatyakar Samaj O Rajniti* (in Bengali), Jugashakti Prakashan, Karimganj.

Gait, E.A. (1981), *A History of Assam*, LBS Pub., Guwahati (First published, 1905).

Guha, Upendra Chandra (2006), *Cacharer Itibritta* (in Bengali), New edition edited by Amalendu Bhattacharyya & Sanjib Deb Laskar, Parul Prakashan, Agartala (First published, 1912).

3

Post-Colonial Land Economy of Barak Valley

—Suparna Roy

The British colonialism in India inducted different land systems in different parts of the country. The pre-colonial land systems underwent great changes during the British rule in India. The entire scenario of the land economy began to change again when India gained her independence as the government had given more emphasis on land reform measures with an objective of heralding an era of egalitarian social justice. Abolition of Zamindari, reform in tenancy laws, ceiling on landholding, etc were the major initiatives undertaken after independence. However, the land economy in British India continues to be a popular field of historical research for a long time. The earlier phase of British rule (till about the Mutiny of 1857) was characterized by the tendency to innovation and the leveling of oriental institutions to the requirements of the most radical theorists in Europe [Ranade, 1916: 265]. This tendency was noticed in the British interference with the traditional land and revenue systems in India. The discussion on land policy was conducted with an appeal, on the one hand, to the principles of western economic theory and, on the other, to the facts relating to the

Indian society in general and the agrarian system in particular as they were perceived by the British administrators. In fact, the pre-mutiny phase of British rule was a very eventful period of modern Indian history not only because the very genesis of what is termed the 'land problem in modern India' is in some important respects traceable to the socio-economic impact of the colonial rule on the traditional agrarian structure and, therefore, this 'problem' continues to be one of the challenging areas of research. The agrarian research in India, as a focal area of scientific social enquiry, was initiated and propelled by the chain of momentous developments in the political sphere during the eighteenth and nineteenth centuries. Political factors also played a primary role in bringing about a shift in the focus of intellectual discourse even during the later period of British rule. The political situation in India was vastly changed after the Mutiny. There was a deep and widespread dissatisfaction among different classes of Indian society which were adversely affected by the impact of British policies, especially land and revenue policies. Land problem and land policy (including revenue policy) were proposed to be crucial problem areas for intellectual enquiry by Indian nationalists. This concern for the agrarian relation and its effects on economic progress was a very important part of the institutional approach upheld by the early nationalists. Among them, Ranade considered institutionalism to be the key for the development of an 'Indian political economy' and questioned the view that "the truths of economic science... are absolutely and demonstrably true and must be accepted as guides of conduct for all time and place whatever might be the stage of national advance" [Ranade, 1916 : 3].

Geographical and Social Settings

The Barak Valley forms the southern part of the state of Assam in India's north eastern region. The valley is surrounded by hills on three sides and has a riverine international border with Bangladesh on the other side. Before independence and partition of India it was a part of the larger Surma Valley (consisting of two districts, namely, Sylhet and Cachar) which was transferred by the British from the Dacca Division of Bengal to Assam in 1874. The state of Assam today is composed of three physical divisions, namely, (i) the Brahmaputra Valley, (ii) the Barak Valley and (iii) the hill range. The two Valleys derive their names from the respective main rivers, the Brahmaputra and the Barak, flowing through east to west in the respective valleys. The hill region (consisting of the Karbi Anglong and the North Cachar Hills districts) separates the two valleys from each other. The Barak Valley region shares its border with the North Cachar Hills district and the state of Meghalaya in the north, the state of Manipur in the east, the state of Mizoram in the south, and the state of Tripura and the Sylhet district of Bangladesh in the west. This plain tract is a geographical extension of the Eastern Bengal plains through Sylhet without any natural barrier intersecting the landscape in between. The valley is predominantly inhabited by the Indo-Aryan population, and the settlements in Cachar plains developed in early times through the immigrant peasants from Bengal in their north-eastern march to the farthest limit of the Indo-Gangetic plains in search of agricultural land. In fact, now divided between India and Bangladesh the Cachar-Sylhet region is a single valley formed by the river Barak. In the British period it was called Surma Valley after a branch of the river, which flanked Sylhet town, and the undivided Valley was the homeland of a common dialect group of Bengali from earliest historical time. [Bhattacharjee, 1996]

Pre-Colonial Land System

The Barak Valley today consists of (i) Cachar plains (present Cachar and Hailakandi districts), which was part of the Dimasa kingdom before 1832 and where Captain Thomas Fisher, the first British Superintendent, introduced changes in the land system, and (ii) Karimganj district, which was a part of old Bengal territory and a part of the Sylhet district till 1947 and was permanently settled ever since Lord Cornwallis .introduced the system in Bengal Prior to the British rule, the land system in Sylhet was as in the rest of Mughal Bengal. On the other hand, the Cachar plains were annexed by the British from the Dimasa kings after the assassination of the last Dimasa king Govindachandra in 1830. The landholding system in the Cachar plains, which centred on the *khel* or principle of voluntary association, was unique to itself. The evolution of the *khel* system from its origin has been historically traced by J B Bhattacharjee in a seminal paper [Bhattacharjee, 1996] which may be summarized as follows : The early settlers in the Cachar plains were almost all peasants. They organized themselves in groups and then proceeded beyond the settled areas to occupy fellow tracts of the sizes that would be enough for the subsistence of the group. They set up the boundary of the area they wanted to develop, reclaimed the tract collectively, started cultivation and established the village in a part of the area. The area reclaimed by a group was known as a *Khel*. The *Khel* was an agricultural guild, collectively managed by the inhabitants. They elected from their rank a headman, called *Mukhtar,* who managed the affairs of the *Khel* with the help of the elders. Every *Khel* had a *Mukhtar* and a *Mukhtar* was locally known as *Bhuiyan* or *Laskar*. Each *Khel* was autonomous or self-governing for all purposes. In this process the *Khels* came into existence one after another till the natural limit of the plains at the foot of the hills was reached. This process of extension of settlements might have continued throughout the early period when the region formed part of the territory under the rulers of Samatata, Kamarupa, Harikela, Vanga and Srihatta in different points of time. The cradle of the settlers was the neighbouring district of Sylhet which was a thickly populated district and where land was an extremely scarce commodity, whereas Cachar had enough fallow land and the population was very small. This factor induced the peasants of Sylhet to extend the cultivation to this tract gradually and to pitch settlements in the reclaimed areas. There was no natural barrier between Cachar and Sylhet, and the Cachar portion of the valley neither experienced the existence of any local state to create a political barrier nor was it then peopled by any other ethnic group. The marshy and swampy valley was indeed uninviting to the shifting cultivators inhabiting the neighbouring hills. In the early British records these *Khels* have been described as agricultural guilds, corporations and commonwealth formed primarily for agriculture. Some *khels* were set up for mining (e.g. salt-manufacturing), fishing and elephant trapping, although these *khels* also had to be affiliated to the *Raj* in which the particular *Khel* was located. The principle of such association was purely mutual convenience, and their existence depended on the voluntary co-operation of the members of the *khels*. They had some obligations to the political authority; firstly, the collective payment of revenue, and secondly, the discharge of some customary services. The latter included the supply of labour to the king. As the number of *khels* increased, the contiguous *khels* voluntarily joined together into larger unions (called *raj*) marking the second stage in the development of the principle of voluntary association in the revenue system of Cachar. The Dimasa kings acknowledged the existence of the *khels* and refrained from interfering in the internal affairs of the *khels*. In fact, a large number of new *khels* came into existence during the Dimasa rule in the valley. [Bhattacharjee, 1996 : 1-3]

Colonial Land Tenure

The British administrators inducted changes in the land system in Cachar by gradually liquidating the *khel* system, by encouraging immigration from the neighbouring districts as part of the colonial policy of revenue maximization, and by declaring the larger portion of the valley as 'waste land' which made it eligible for 'tea grants' under 'waste land rules'. However, Thomas Fisher, the first Superintendent, continued the native revenue system, which he found in force for a very long time and the interference with which, he thought, might affect the sentiment of the people. He abolished duties on the river *ghats,* except a designated few on the Surma at which duties were levied on exportable commodities at moderate rates. The export of agricultural products was totally exempted from duty. The state monopolies on betel nut, timber, salt, cotton, grass etc., which prevailed since Dimasa rule, were abolished, retaining the hill *ghats* which offered the only means of raising revenue from the dwellers of the adjoining hills. He raised the revenue for the government by inducing and giving settlement to a large number of immigrants from Sylhet and the neighboring districts. But P. Mainwarring, who succeeded Fisher in 1835, abolished the *Mukhtar* system as he found this system to be expensive and inefficient. J.G. Burns, the next Superintendent of Cachar, completely liquidated the *khel* system by effecting a settlement in 1838 with the occupants and settlers for five years. Burns also took initiative in attracting new settlers and during his tenure large tracts of fallow and forested lands were brought under assessment. Major changes were introduced by the next two Superintendents, namely, G Verner and R Stewart, who effected new settlements in 1843-44 and 1858-59 respectively enforcing ryotwari system in Cachar. It was during this period that a large number of tea grants were awarded to European planters. As a result of these settlements which remained in force for the rest of the colonial period, at least five types of settlements coexisted in Cachar within the basic framework of the ryotwari system; viz. (1) the *Mirasdari* tenure of Stewart's settlement, (2) the grants under the Old Assam Rules, (3) the grants under Fee Simple Rules, (4) the leases of waste land under the old and new *Mirasdari* Rules, and (5) the revenue-free tenures such as *baksha* and *Lakhiraj*. [Barpujari, 1963 : 90; Bhattacharjee, 1977 :80; Baden Powell,1990 :437; Guha, 1977 :47; *The Assam Land and Revenue Regulation, 1886*] Evidently, immediately after independence the Barak Valley region had two types of land settlements in two different parts, namely, ryotwari settlements in Silchar and Hailakandi subdivisions and permanent settlement in Karimganj subdivision, which called for reforms in conformity with the principles of social justice enjoined by the constitution of the new nation.

Post-Colonial Land Reform Measures

Land tenure system has its importance as it locates the owner of land, and influences the productivity of land and the social organization prevalent in the state or region. Like other states of India, the land tenure system in Assam (including Barak Valley) was found to be unsatisfactory and it was decidedly unfriendly to the tenants. The Zamindary system prevalent in the region was almost similar to the system prevalent in Bengal and Uttar Pradesh. Under the system one or more persons owned a village and were responsible for the payment of land revenue. There were permanent settlements in the Karimganj Subdivision (now district). The system was not at all beneficial due to the following reasons: (i) it robbed the state of its due share, (ii) made the landlords absentee parasites, proved to be harmful to the cultivators, and stood on the way of agricultural development, (iii) discouraged enterprise on the part of cultivators due to frequent enhancement of rent and fear of ejection, and (iv) the agricultural sector suffered as the zamindars contributed nothing for agricultural progress. These evils justified the abolition of the zamindary system to bring the huge number of tenants into direct relationship with the state. The main objective of land reforms, however, was to alter the pattern of land ownership and to bring a structural change in rural society for the attainment of greater production and social justice. Accordingly, in the approach paper of the second five year plan of India it was mentioned that the objective of land reform was to create conditions for evolving the agrarian economy with high degree of efficiency and productivity and to establish an egalitarian society as early as possible. [Goswami, 1988]

The land reform measures were also introduced in Assam along with rest of the country . The basic motives of various land reform measures initiated in Assam were:

1. Providing security of tenancy right to the tenants;
2. Protecting the tenants from undue exploitation by the land owners;
3. Establishment of a direct nexus between the tenants and the government;
4. Fixation of ceiling on land holdings; etc.

It is to be noted that the land reform measures in India were framed on the basis of certain principles. These principles were laid down by the Agrarian Reforms Committee (1948) and accepted by the Planning Commission. These are:

1. The agrarian economy should develop the farmer's personality;
2. There should be no scope for class exploitation;
3. There must be maximum efficiency of production; and
4. The scheme of reforms should be practicable.

The above principles involved the following major changes: (a) abolition of zamindari, (b) reform of the condition of the tenants through legislation, (c) an increase in the unit of cultivation, (d) consolidation of holdings, and (e) adoption of mechanized cultivation. [Goswami, 1988]

The government of Assam introduced various land reform measures in the state since independence. The main purpose of such land reforms was to alter the pattern of landownership and to bring about such structural changes in the rural society as to ensure that the fruits of production are equitably enjoyed by those who toil in the agricultural land. The tenancy reforms legislations in Assam, therefore, envisaged providing fixity of tenure, curbing the freedom of landowners to eject tenants, and fixing the minimum rate of rent. In fact, the first step towards tenancy reforms in Assam can be traced back to 1929 when the Goalpara Tenancy Act. 1929 was enacted. This Act covered the Goalpara district (a former Bengal territory and permanently settled) alone, but it had to be enacted under nationalist pressure and it marked an important milestone in the history of land reforms in Assam. This was followed by the Assam Temporarily Settled Districts Tenancy Act (1935) to protect the tenants under the landholders of the Brahmaputra valley and Cachar, and the Sylhet Tenancy Act (1936) to protect those in the permanently settled areas of Sylhet (including Karimganj). These Acts aimed at improving the conditions of the tenant class by conferring upon them occupancy rights - permanent, heritable and transferable. They also conferred the rights of use and occupancy on the sub-tenants and under-tenants and accorded protection to the tenants against illegal ejection and enhancement of rent. These Acts remained in force even after the abolition of the Zamindari system in 1956-57 and was amended in 1970 to include the provisions of Adhiar Tenant Protection Act, 1948. According to the provisions of the Adhiar Act the government was required to establish Adhiar Conciliation Boards in specified areas, each Board having one representative of the landlords and another from the Adhiars (i.e share croppers) with the Revenue Officer of the area as the chairman to ensure that the provisions of the Act are followed and that there is no illegal eviction of an Adhiar. It was also to settle any dispute arising between a landlord and an Adhiar, and its direction was to be executed by the Revenue Officer. From the gross produce of crops, an

Adhiar was required to return at first the seed grains, if any, supplied by the landlord or any other person under whom the Adhiar holds the land. The balance of the produce, thereafter, is to be divided between the landlord and the tenant as follows: (i) Where the landlord supplies the plough-cattles and the cultivation is done with their help, the landlord will get one-quarter share of the produce and the tenant three-quarters; (ii) Where the landlord does not supply the plough-cattles, he will get one-fifth of the produce and the tenant four-fifths. But if the contract rent is lower than the above share, the landlord will only receive the contract rent. When more than one crop is grown, the landlord will get his specified share only of the major crop. A tenant has the option to pay the money value of the crop rent instead of the produce.

As a matter of fact, the Act of 1948 failed to provide security to the sharecroppers. Therefore, the Assam (Temporarily Settled Areas) Tenancy Act 1971 came into force with effect from 10th December, 1971, initially in all the temporarily settled areas of the plains districts, by repealing the Tenancy Act of 1935 and the Adhiar Protection Act of 1948. The Landholding Act of 1974 repealed the Goalpara Tenancy Act, 1929 and the Sylhet Tenancy Act, 1936 and enforced the Tenancy Act of 1971 in the formerly permanently settled areas of Goalpara and Karimganj. This Tenancy Act of 1971 is now applicable in the entire state of Assam, except two hill districts, viz. North Cachar Hills and Karbi Anglong. [Dhar, 1997]

Along with other states of India, the Zamindari system in Assam has also been abolished. Two Acts have been passed since independence for the abolition of intermediaries in the permanently settled estates. These are (a) The Assam Management of Estates Act (introduced in 1949 and enforced in 1952-53) and (b) The Assam State Acquisition of Zamindaries Act (introduced in 1948 and enforced in 1953-54). Apprehending the abolition of zamindari system after independence, the zamindars started the reckless exploitation of forests, fisheries and other natural resources, and also began to lease their lands with under-ryots at heavy premiums. The Assam Management of Estates Act empowered the government to prevent (and if necessary, even to take over the management of the land) such exploitations. Because of the delay in enforcement of the Act due to the legal proceedings, it is doubtful whether the state has been able to prevent hasty and reckless exploitation by the zamindars. In March 1948, a bill to abolish the zamindari system in the state was presented in the state assembly which became an Act in 1951. This Act extends to all the zamindari estates of Assam, to the Lakhiraj estates within the boundaries of the zamindari estates and to acknowledged estates. The Act enables the government to acquire all the interests associated with these estates, like right to minerals, forests, fisheries, realization of rents and other incorporeal rights. The state is to acquire the estates free from all encumbrances, and the existing encumbrances are to be deducted from the compensation payable to the zamindars. It is also provided that the tenants in these estates shall continue to hold land on the same conditions or terms as were in force immediately before the date of enforcement, until further legislation is passed by the state legislature. [Joshi, 1975; Dhar, 1997]

The Act provided that every proprietor or tenure holder will get compensation at the prescribed rates for the land taken away from him. It also provided that the proprietor, i.e 'a person holding in trust or owning for his own benefit any permanently settled estate or a part of any such estate' or the tenant-holder i.e 'a person who has acquired from a proprietor or from another tenure-holder right to hold land for the purpose of collecting rents or bringing it under cultivation or by establishing tenants on it' will retain possession of the following:

1. The homestead in his possession on the date of vesting.
2. Buildings, together with lands on which they stand, other than buildings which are used for purposes of trade, commerce or agriculture, or are let out on hire or rent, and are in his possession on the date of vesting.
3. Private land up to 400 bighas. The Government may relax this limit in the case of large scale farming on a co-operative basis, or where power driven mechanical appliances are used. Such private lands of the proprietors or intermediaries will be assessed, according to the rules of the Temporarily Settled Estates, after the abolition of zamindari.
4. Tea gardens or Orchards. Land held for cultivation of tea shall continue to be held under the same terms and conditions as existed before the date of vesting. [Gangopadhyay, 1990]

The Assam Consolidation of Holdings Act, 1960 was enacted in 1961 for consolidation and prevention of fragmentation of agricultural holdings for better cultivation in the plain districts of the state. The scheme was originally taken up to save labour and money in agricultural operations and also to increase agricultural production. Another very important land reform measure implemented in Assam was the land ceiling. The Assam Fixation of Ceiling on Land Holdings Act was passed in 1955. The Act provided for fixation of ceiling on land holdings at 150 bighas per family. Later on, it was reduced to 75 bighas by making necessary amendment. It was further amended in 1972 to limit the amount of agricultural land that can be held by a family to 50 bighas and to reduce the allowable area of orchards from 30 bighas to 15 bighas. In order to protect the interest of the plains tribe people of Assam, the Assam Land and Revenue Regulation (Amendment) Act, 1947 was passed under which 30 tribal belts were constituted in the areas predominantly inhabited by the tribals in the plain districts of Brahmaputra Valley. [Goswami, 1986]

Conclusion

The economic condition of Cachar before the advent of the British was traditional and undeveloped. There was a political disorder in the beginning of the nineteenth century as the authority of the Dimasa king was challenged internally and externally. The fugitive princes from Manipur - Marjit, Chaurjit and Gambhir Singh – scrambled to grab the reign. This was followed by the Burmese invasion, the declaration of war by the British against Burma, the Treaty of Badarpur between Gobindachandra and the British, the First Anglo-Burmese War, and signing of the Treaty of Yandaboo in February 1826 between the British Government and the King of Burma. Under its provisions Cachar was restored to Gobindachandra who agreed to pay Rs. 10,000 annually to the British as the ruler of a protected state. Allured by the vast economic resources of Cachar, the British began to fish in the troubled water in Cachar, the revenue of which was then calculated at rupees three lakhs. The assassination of Govindachandra offered a golden opportunity to the British to annex Cachar by a Proclamation in August 1832. After the annexation of Cachar, the Supreme Government appointed Lieutenant Fisher as its Superintendent with the headquarters at Dudpatil. However, in pursuance of the expressed desire of the Governor General, Lord William Bentinck, Cachar was placed under the control of the Commissioner of Dacca. The headquarters of the Superintendent of Cachar was shifted to Silchar in 1836. The new administration found that in addition to political chaos, the valley was liable to severe floods every year. Cachar was depopulated during the Manipuri devastation and the Burmese wars. Following the assassination of Raja Gobindachandra in 1830 many people fled away to Sylhet and other safe places. The period between 1800 and 1830 was indeed remarkable for large scale depopulation in Cachar. J.B. Bhattacharjee observed that towards the end of the rule of Raja Krishna Chandra (1790-1813) the economic structure of the Dimasa Government was in a decaying state and several new taxes and demands had to be imposed on the people to meet the crisis. [Bhattacharjee, 1977] The advent of the British was followed by radical changes in revenue administration. Land was identified as the most important wealth and from the very beginning the administrators were interested in increasing the revenue. It may be understood from the fact that at the time of annexation of Cachar (1832) the settled area was 20,663 acres yielding a revenue of Rupees 21,411, whereas by 1918 it rose to Rupees 6,99,964. As early as in 1872, the demand on account of land revenue amounted to Rupees 1, 53,345 which increased to Rupees 7 lakhs on the eve of the departure of the British from India. The most remarkable economic enterprise during the period was the tea industry. In order to encourage tea cultivation the government granted waste lands to the planters in the form of revenue-free or nominally revenue- paying estates in perpetuity. The Waste Land Settlement Rules were first framed in Assam in 1838, but these were modified in 1860, 1864, 1875, 1876 and 1882 to expand the scope to include the forested, fallow and abandoned land for tea grants. In fact, very large areas of land were leased out in Cachar for tea cultivation under these rules, although only lesser parts of the grants were actually utilized for tea. The tea planters emerged as big zamindars in the district; they brought labour from outside and used the land for various other purposes. There were complaints of exploitations in the tea estates. No wonder, they became the targets of peasant and labour movements which were inspired by the leaders of the national movement. [Bhattacharjee, 1977]

The scenario gradually changed after Independence under democratic demand for land reforms. The legislations were introduced from time to time, but implementation suffered due to corruption and delay. The main tasks such as preparation of the records of rights, calculation of the compensation, settlement of land with tenants, measurement and distribution of excess land, fixation of Adhiars' shares etc., were vested on low-paid subordinate revenue officers like Mandals and Kanongos who are generally susceptible to temptation and indifferent to accuracy and promptness. Though land reform measures have been enacted several years ago, the progress of implementation has been very slow. Therefore, it has become essential to appoint a high level committee to examine the progress of implementation of land reform measures, their impact on land management and crop husbandry, defects in the Acts, impact on land productivity, etc. It is also doubtful whether the present legislation, even when fully implemented, will be able to stop exploitation by large landowners without creating difficulties for small landholders. A rethinking of the whole of the land reform policy is necessary so that the reform measures can achieve an overall economic development by encouraging larger and more efficient agricultural output and a better distribution of the returns of agriculture.

REFERENCES

Barpujari, H K (1963), *Assam in the Days of the Company*, LBS, Gauhati.

Baden Powell, B.H. (1990), *The Land System of British India*, Vol. III, Reprint D K Publishers, New Delhi.

Bhattacharjee, J B (1996) "Khel : Roots of Agricultural Commonwealth in Cachar", *Journal of Assam University*_Vol.1 No.1, January 1996.

Bhattacharjee, J B (1977), *Cachar Under British Rule in North East India*, Radiant Publishers, New Delhi.

Dhar, P.K.(1997)., *The Economy of Assam*, Kalyani Publishers, New Delhi.

Gangopadhyay, D.K (1990). *Revenue Administration in Assam*, Revenue Department, Government of Assam, Dispur.

Goswami, Atul ed. (1986), *Land Reforms and Peasant Movements*, United Publishers, Guwahati.

Goswami, P.C.(1988), *The Economic Development of Assam*, Kalyani Publishers, New Delhi.

Govt. of Assam (1886), *The Assam Land and Revenue Regulation, 1886*, Shillong.

Guha, Amalendu (1977), *Planters Raj to Swaraj, Freedom Struggle and Electoral Politics in Assam 1826-1947*, ICHR, New Delhi, 1977.

Joshi, P.C. (1975)., *Land Reforms in India*, Allied Publishers, New Delhi.

Ranade, Mahadev Govind (1916), *Essays on Indian Economics*, G.A. Nateson & Co., Madras.

4

Disparities in Economic Development: A District Level Study of Assam with Special Reference to Barak Valley

—Sumanash Dutta

The state of Assam is a land of hills and plains. It is comprised of three well defined regions - the Brahmaputra plains in the north, the Karbi Anglong and North Cachar Hills in the middle and the Barak plains in the south. The mainstay of the economy of Assam is agriculture with more than 70 percent of the population engaged in it. The principal crops of the state are rice, maize, wheat, pulses, oilseeds and potatoes. Different reports published from time to time (e.g. the Assam Development Report, 2002; East India Human Development Report, 2004; Economic Survey, Assam, 2006-07) present a dismal picture of the socio-economic condition of Assam in general and of incidence of poverty in the state in particular. Between 1951 to 1979, Assam's GDP grew more or less at the same rate as that of India as a whole, but since then Assam has been continuously slipping in the process of economic growth compared to the other states. As per Planning Commission estimates, 36.1 percent of the state's population live below the poverty line, compared to the all-India average of 26.1 percent in the year 1999-2000 [*Statistical Hand Book of Assam, 2006*]. The State Development Report (2002) has confirmed that poverty in Assam has been primarily a rural phenomenon as 40 percent of the rural people are in the BPL category, whereas in urban areas it is as low as at 6.29 percent. The District-wise poverty statistics quoted in the Report reveal that the western plains, the flood prone and thickly populated south-east plains and the tribal areas are the poorest. In an average, a person living in a state like Punjab earns two and half times more than a person living in Assam. The Report points towards disparities of economic development in the state of Assam. In the demographic front, the density of population has been increasing over time with galloping strides. More importantly, there are some alarming zones, such as Nagaon, Dhubri, Kamrup, where density of population is as high as near about 600 as in 2001.

As per the East India Human Development Report prepared by NCAER (2004) Assam is distinctly laggard in terms of important indicators of socio-economic well-being even among the states of Northeast India. In the urban areas of Northeast India, the monthly per capita consumption expenditure is lowest in Assam, at Rs.169.93 only. In rural areas also, it is lowest in Assam, at Rs.153.6 only. The level is so low that it does not allow any disparity, but subsistence only. Although in case of effective literacy the state of Assam is ahead of the national average, the gender disparity in literacy rate in Assam is as high as 15.9 percent. The rural-urban disparity in respect of literacy is also quite pronounced in Assam. In the Northeast, the drop out rate is high in Arunachal Pradesh, Assam, Tripura and Manipur. The expectation of life at birth in Assam of males and females based on the 1980 mortality rate (as revealed by the SRS) is 58.7 years and 58.5 years respectively. This is worse than that of India for both males and females. Men live longer than women in Assam, though for the country as a whole the women live longer than men. In regard to health care services, the condition of Assam is worse than rest of the states in the Northeast. As against the all India figure of 0.73 hospital beds for every 1000 persons, Assam has only 0.06 hospital beds per thousand people. The size of the population served per Doctor is the highest in Assam among north-eastern states. The state has one hospital for 107,000 persons. Assam has also the lowest number of Family Welfare Centres per one lakh population, i.e. at 0.6. Data relating to the total dependent population reveal that the size of dependent population is the second highest in Assam (84.2 percent) among the north-eastern states. The work participation rate is also lowest (at 36.5 percent) in Assam.

Assam is distinctly a laggard among various states of India. Among fifteen major states of India, the rank of Assam, as per life expectancy index, is 14th with life expectancy index value of 0.37 [The Economic Survey of India, 1997-1998]. In terms of the educational index, the rank of Assam is 6th with educational attainment index value of 0.4187 [CMIE, 1996; The Economic Survey of India,1995-1996]. The per capita real SDP index of Assam is 0.28 and its rank out of fifteen major states of India is 8th [Indian Public Finance Statistics,1995].

The economy of Barak Valley, a constituent part of Assam, is predominantly agrarian in character. More than 58 percent of the total working population in the valley are either cultivators or agricultural labourers and 70.7 percent of its workers earn their livelihood from the primary sector activities [Census Report,1991]. The agriculture is over crowded and the average yield of rice per hectare in the valley is as low as 1447 kg. as against India's figure of 1745 kg .[Bhattacharjee & Nayak, 2003]. Significantly, the corresponding figure of the Brahmaputra Valley, the most fertile region of Assam, is only 1131 kg, which is lower than that of the Barak Valley. The industrial base of Barak valley has also remained extremely narrow even after 60 years of independence .Out of 17 industrial estates in Assam with 320 functioning sheds, only one estate is in Barak Valley with only 22 functioning sheds [*Statistical Hand Book, Assam, 2006*]. The number of small scale industries registered under the Directorate of Industries and Commerce in Assam is 50694 as on December 2005, and out of these, the share of Barak Valley is only 3767. Tea plantation is the prime organized industry of Barak Valley but this sector is also in a state of sickness and neglect. This is mainly because of low investment out of the surplus earned by the tea planters [Bhattacharjee & Nayak, 2003]

The above discussion, although extremely sketchy, reveals the socio-economic backwardness of the state of Assam within which the condition of Barak Valley appears to be the worse. The apparently existing regional imbalances in economic growth and development in the state may jeopardize the very objective of attaining equitable distribution of social and economic welfare as it does not merely depend upon the level of development but also on its distributive aspect. While studies on the problems and the prospects of Assam in attaining economic development are welcome exercises, the present situation also calls for micro level investigations for deeper understanding of the problem of disparities in economic development in Assam. This presentation is an attempt to examine the level of economic development at the district level of the state and thereby to shed light on the disparities in the economic development of the districts of Assam. The first part of the presentation discusses the overall socio-economic condition of the state with special reference to the Barak Valley of the state. The second part shall attempt to define the term 'regional disparities' and its implications in both theoretical and empirical framework. The third part shall focus on the design of the study and the fourth part will be devoted to the estimation of the result and its analysis, interpretation and conclusion.

II

The theme of regional disparities has been well-researched both theoretically and empirically. Gunnar Myrdal [1957] explained regional disparities at national and international plain with the help of Backwash Effect and Spread Effect. The term region in his analysis refers to units of geographical areas exhibiting more or less same level of economic, social and cultural development both at national and international level. Therefore, regional disparities mean the disparities that exist among different regions. The regions may be as small as administrative units such as districts within a state. On the basis of the level of development achieved, the districts can, however, be lumped together to identify regions showing High, Medium and Low growth rates as per objectives laid down. [Ryngynga, 2003] The size of a region may vary depending on what type of disparity the study attempts to focus on.

The term regional disparity is wide in scope and application .It may explain disparities prevailing in different fronts such as political, social, cultural and economic. Among these, political and cultural disparities are not very popular terminology in the literature of regional disparities. These are rather termed as diversities and very often construed as strength rather than weakness. Coexistence of different political and cultural systems in regions and countries make the world a more attractive place to live in. However, political and cultural diversities have significant influences on human development which are often explained in terms of inequalities in political freedom, opportunities and predicaments. Questions such as 'How will the new constitution of Iraq satisfy demands for fair representation for Shiites and Kurds?' and 'Will the French legislature approve the proposal to ban headscarves and other religious and cultural symbols in public school?' aught to have strong bearings on human development. A study on understanding human welfare under Chinese political system of governance and that of India's will definitely focus on political disparities prevailing in these regions. Similarly, a study on Ahoms of Assam and Tripuris of Tripura can not afford to ignore the cultural disparities that exist between the two communities.

There is no denying the fact that the existing literature on regional disparities broadly refers to disparities in economic development. There is plethora of studies on regional disparities in economic development which took into consideration different indicators of economic development. Selecting eight agricultural indicators, twelve industrial indicators and six indicators of service sector, Kantawala and Rao [1992]) analyzed the inter-state disparities in Sectoral development in India. The study found that over the period from 1970-71 to 1985-86, the relative position of states in India have not changed significantly, considering the overall development. Rangacharyulu [1993]) examined the rural development of 17 major states in India by selecting five core indicators of rural development. The study observed that there was no significant change in the position of majority of the states in rural development over the period from 1971 to 1987. Subramanian [1995] examined the inter-state differences in the level of family health of all the states of India except Tripura. On the basis of composite index of family health, Subramanian classified all states into three categories—High, Low and Medium ranked. The inter-state disparities on the basis of large number of social, economic and infrastructural factors were studied by Rana [1997].The study found the pattern of development of Indian economy was not in unison with respect to all the indicators rather it was of mixed nature. Bhatia [1999] highlighted inter-state variations in rural infrastructural development and emphasized the need for increased investment in rural infrastructure from the point of view of increasing agricultural growth and productivity. Gulati [1999] in his study included 17 major states of India to examine regional disparities in economic development. The findings indicated that the inter-state disparities had widened during the study period of 1960-1991. In his study of regional disparities in north-east region, Dutta [2003] observed that the states of north-east region are suffering from paucity of funds and the states are far behind the other states of India in basic amenities of life.

Since economic welfare does not merely depend upon the level of development but also on its distributive aspect, disparities in economic development has always remained as a matter of great concern for the social scientists. Different studies conducted at different point of time in India not only have indicated the existence of strong economic disparities among the regions, a good number of studies have also found that the degree of disparities have been widening over time making poor regions poorer in relative term. As identification of poorer regions is an important part of development planning, the present paper attempts to find poorer regions in terms of some selected indicators within the state of Assam by applying simple methodology.

III

In order to understand the inter-regional disparities of economic development of Assam, the present study considers the districts of Assam as units of study. Several variables are taken into consideration as indicators of economic development. The selected variables are:

(a) Percentage of Villages Electrified (X1)

(b) Per Capita Gross Domestic District Product in Rs.(X2)

(c) Percentage of Habitations having drinking water facilities (X3)

(d) Literacy Rate in percentage (X4)

(e) Number of Medical Centers [Hospitals, PHC, Dispensaries] (X5}

(f) PWD Road length in Km per one 100 Sq.Km (X6)

(g) Sex ratio (X7)

(h) Percentage of Main Workers to total workers (X8)

Data pertaining to the selected variables are collected for each district of Assam with reference year mainly of 2001. However, data of couple of variables are of 2005-06, as data for these variables of 2001 were not readily available. Variables have been selected under guiding principle that higher values of these would mean higher economic development in a restricted sense. The technique Factor Analysis is used to derive the relative weight of selected indicators of economic development. With the help of the relative weights of the indicators, thus estimated, a composite index of economic development is constructed for each District of Assam to analyze the inter-district variations or disparity in economic development.

IV

The Mean and Standard Deviation of the selected variables are shown in Table-1.

The high S.D values of variables X1,X2,X5,X6 and X7 indicate inter-district disparities in attainments in the field of percentage of villages electrified , per capita GDDP, health care, PWD road and sex ratio. The results of Factor Analýsis and derived objective weights of the selected indicators are presented in Table-2.

Table-1

Mean and Standard Deviation of Selected Variables

Variables	*Mean*	*S.D*
X1	73.73913	21.36313
X2	11628.7	4992.968
X3	89.72435	9.886022
X4	63.11652	7.342276
X5	38.52174	16.06434
X6	46.47826	17.91107
X7	934.7391	16.61111
X8	0.750435	0.059274

Table-2

Factor Analysis Results and Weights of the Indicators

Variables	*Factor 1*	*Factor 2*	*Factor 3*	*Factor 4*	*Communality*	*Weight*	*% Weight*
X1	o.128	0.615	0.642	0.245	0.867	0.11342	10.8693
X2	0.793	0.07855	0.02733	0.207	0.678	0.217076	20.8028
X3	-0.570	0.393	-0.088	0.613	0.863	0.09338	8.9488
X4	0.669	0.184	-0.376	0.431	0.809	0.183132	17.5499
X5	0.441	0.540	0.474	-0.290	0.796	0.111467	10.6821
X6	0.0932	0.506	-0.372	-0.677	0.861	0.104449	10.0096
X7	-0.752	0.472	0.05299	-0.0697	0.796	0.09743	9.3369
X8	-0.0533	-0.555	0.697	-0.0816	0.803	0.123139	11.8007
Var Explained	27.374	20.642	17.677	15.234			
Cumm. Var Explained	27.374	48.016	65-682	80.916	Total	1.043493	100

With the weights of the variables and their standardized values, the composite index of economic development is estimated District-wise. The composite index values District wise are shown in Table-3.

Table-3
Composite Index and Rank of the District

District	*Composite Index of Economic Dev*	*Rank*
Dhubri	10.18487	14
Kokrajhar	10.0118	22
Bongaigaon	10.51765	10
Goalpara	10.16432	16
Barpeta	10.44451	11
Nalbari	10.77852	4
Kamrup	11.24822	2
Darrang	10.34531	13
Sonitpur	10.12627	18
Lakhimpur	10.1675	15
Dhemaji	10.05933	20
Morigoan	10.76598	6
Nagaon	11.04572	3
Golaghat	10.69193	7
Jorhat	10.7747	5
Sibsagar	11.24949	1
Dibrugarh	10.62938	8
Tinsukia	10.01859	21
KarbiAnglong	10.13353	17
NC Hills	9.86633	23
Karimganj	10.37071	12
Hailakandi	10.08783	19
Cachar	10.54017	9

In Table-3 all districts of Assam are ranked according to the composite index value .The rank of Sibsagar district is 1st and the last rank (23rd) is that of N.C Hills. In Barak Valley, the condition of Hailakandi district is the worse. The position of this district in Assam is 19th .The position of Cachar district is slightly ahead of Karimganj .However, the overall condition of the Barak Valley strengthens apprehension that the valley is the most backward region of the socio-economically backward state of Assam. The disparities among the districts according to economic development defined only by the selected variables, is low as the values of the composite index ranges between a small interval of 9.87 to 11.25 (corrected up to second decimal places).The result further reveals that the exhibited disparities is mostly explained by a single indicator, namely, the per capita Gross Domestic District Product which alone carries near about 21 percent weight in defining the level of economic development. The next important variable is literacy which enjoys near about 18 percent weight in defining economic development. In respect of other variables such as percentage of villages electrified, percentage of habitations having drinking water facility, health care, availability of PWD roads, sex ratio and percentage of main workers to total workers, there exist very little differences among the districts of Assam. The overall low level of disparity in economic development is reflected in low value of Standard Deviation of composite index values of economic development which is only 0.399072.

To gain further insight into the problem of regional imbalances, the average composite index values of three regions of Assam namely the Brahmaputra Valley, the Barak Valley and the area comprising two hill districts of N.C Hills and Karbi Anglong are estimated .The estimated values are shown in Table-4.

Table-4
Average Composite Index Values of Three Regions

Region	*Average Composite Index Value*
Brahmaputra Valley	10.51246
Barak Valley	10.3329
N.C Hills & Karbi Anglong	9.9999
S.D	0.260059

From Table-4, it appears that the region wise disparity is less in Assam than the district wise disparities in economic development. However, the picture is clearer now and Table-4 shows that Brahmaputra Valley is ahead of Barak Valley and N.C Hills, Karbi Anglong region of the state. The figures show that in terms of the indicators of economic development, Karbi Anglong and N.C. Hills region is the poorest region of the state.

Conclusion

Many development and growth theories have proved that regional disparities and imbalances are inherent in the progress of development and their degree goes on changing with the stages of economic development. The economy of the state of Assam exhibits some direct symptoms of illness which no one can even pretend to ignore .Among the fifteen major states of India, the socio-economic condition of Assam is precariously low. The extremely low level of socio-economic development has embodied a very low level of disparities among the districts and regions of the state in the concerned direction. The immediate need of the hour is to launch a concerted effort involving all sections of the population to achieve faster economic development. However, the present study points out that the seed of regional and district level disparities is very much present even in the low level of economic development of Assam which is undoubtedly an area of concern. The faster growth rate of economic development may aggravate the situation unless it is equitable in nature.

REFERENCES

Bhatia, M.S. (1999), "Rural Infrastucture and Growth in Agriculture", *EPW*, March 27 to April 2, Mumbai.

Bhattacharjee, P.R & P. Nayak (2003), "Socio-economic Rationale of a Regional Development Council for the Barak Valley", *Journal of North East India Council for Social Science Research*, Vol. 27, No. 1.

Dutta,P.C. (2003), "Regional Disparities and Its Consequences in North Eastern Region of India", in *Economic Liberalization and Rural Disparities in India* edited by A.C Mahapatra & C.R Pathak, Star Publishing House, Shillong.

Gulati, R.K. (1999), *Regional Disparities in Economic Development*, Deep and Deep Publications Pvt. Ltd, New Delhi.

Kantawala and Rao (1992), "Sectoral Development in India—An Inter-state Analysis", *Indian Journal of Regional Science*, Vol. XXIV, No.1.

Myrdal, Gunnar (1957), *Economic Theory and Underdeveloped Regions*, University of Stockholm, Vora & Co Publishers Pvt. Ltd.

Rangacharyulu,S.V (1993), "Composite Development Index of Rural Areas", *Journal of Rural Development*,Vol.12(1).

Rana, R.K (1997), "A Study of Inter-state Disparities and Pattern of Development in India", *Indian Economic Association conference volume*, Hyderabad.

Ryngynga, Kamtibon (2003), "Levels of Development in the Hills of North-East India" in A.C Mahapatra & C.R Pathak (ed) *Economic Liberalization and Rural Disparities in India*. Star Publishing House, Shillong.

Subramaniam, M.S (1995), "Inter-state Variation in the Development of Family Health in India - A Taxonomic Approach", *Indian Journal of Regional Science*, Vol.XXVII, Nos.1 & 2.

5

Exploring Economy of Barak Valley for Development Strategy

—Niranjan Roy

The state of Assam is comprised of three physical divisions, namely, the Brahmaputra Valley, the Barak Valley and the Hill range. The Brahmaputra Valley, which forms northern part, is the largest division comprising 71.7 percent of total geographical area of the state. On the other hand, the Barak Valley region, which forms the southern part, is comparatively smaller in size. The two valleys derived their names from the respective main rivers, the Brahmaputra and the Barak flowing through east to west in the respective valleys. The hills division formed by the Karbi Anglong and the North-Cachar hills stands between the two valleys.The Barak Valley region is situated between Longitude 92^{0} 15′ and 93^{0} 15′ East and Latitude 24^{0} 8′ and 25^{0} 8′ North covering an area of 6922 sq. km. The valley constitutes 8.9 percent of the geographical area of the state but it contains 11.22 percent of the population as per 2001 census. The region shares its border with North Cachar hills district and the state of Meghalaya in the north; the state of Manipur in the east; the state of Mizoram in the south and the state of Tripura and the Sylhet district of Bangladesh in the west.

Administratively, it is comprised of three districts, namely, Cachar, Karimganj and Hailakandi. The Cachar is the largest and the Hailakandi is the smallest district with total geographical areas of 3786 sq. km. and 1327 sq. km respectively, and that of Karimganj district is 1809 sq. km. The district headquarters of Cachar, Karimganj and Hailakandi are Silchar, Karimganj and Hailakandi respectively. Of the three districts only Cachar district is comprised of two sub-divisions, namely, Silchar and Lakhipur, and other two districts are without sub-divisions.

The Barak Valley region has an undulating topography characterized by hills, hillocks (locally known as *tillah*), wide plains and low-lying waterlogged areas (locally called *beel*). Most of the hills have a north-south spread interspersed by strips of plains. The principal river of the region, namely, the Barak takes its rise a little to the west of Maothana, on the northern boundary of Manipur. Taking a south-west turn from its source near Tipaimukh in Manipur, it then comes to north and for a considerable distance forms the boundary line between erstwhile undivided Cachar and the state of Manipur. After its junction with the river Jiri, it turns again to the west and flowing through the heart of Cachar district reaches Badarpur in the Karimganj district. From Badarpur to Haritikar, it provides the boundary between the Karimganj district and Bangladesh. At Haritikar, the river is divided into two branches, namely, Surma and Kushiara. The former enters Sylhet (Bangladesh) at the junction, while Kusiara flows to Karimganj town forming India's boundary with Bangladesh and thereafter, this branch too enters Bangladesh. In addition to Jiri, the Barak receives numerous tributaries from the neighbouring hills, like Chiri, Barail, Madhura, Jatinga, Dalu etc. The other important rivers of the region are Sonai, Daleswari, Katakhal, Longai, Singla and Kakra. Most of the rivers dry up during winters. These rivers are not navigable even in the rainy season, although most of them rise up in high spate during summers and cause floods almost every year in the region [Bhattacharjee, 1977].

Cachar and Sylhet districts were transferred in 1874 by the then British Government from the Dacca division of Bengal to form together the Surma Valley division of Assam. The partition of India in 1947 was the partition of Assam, partition of Surma Valley and partition of Sylhet as well. Sylhet was handed over to East Pakistan (now Bangladesh) leaving to Assam only three and half *thanas* (which now constitute the Karimganj district). Karimganj, Silchar and Hailakandi districts (all subdivisions of the erstwhile Cachar district) constitute the region now known as Barak Valley. As a result of the partition, the Barak Valley lost its vital road and waterway links with the rest of India which were always through East Bengal. Though the rail and road connections were restored through the hill range and the Brahmaputra Valley after independence, the communication through these mountainous and longer routes has proved to be difficult, prone to frequent disruptions, and a high cost affair. The tea industry, which was once a pride of the valley, is on decline since the partition.

Climatic Conditions

The climate of the Barak Valley region is subtropical, warm and humid. The average rainfall of the region is 3180 mm with average rainy days of 146 per annum (as per estimate based on data recorded by the Regional Agricultural Research Station, Karimganj). The region owes this high rainfall mainly to south-west monsoon, which usually operates for a longer spell in the Northeast region compared to other parts of India. The monsoon rains normally starts from early June and continues to October. Even pre-monsoon heavy showers in late March and April are not uncommon. During the summer months temperature generally varies between 25^0 and 40^0 Celsius, while during the winter season the temperature ranges between 10^0 and 25^0 Celsius. The humidity remains high throughout the year with minor recess during the months of March and April. The relative humidity varies from 58 percent to 91 percent. A consequence of heavy rainfall during the summer months is the widespread water logging and several rounds of floods in the region. The floods cause enormous loss of life, crop and property every year.

Land-use Pattern

The pattern of land-use of a country is determined by the physical, economic and institutional framework taken together. The existing land-use pattern in different regions in India has been evolved as the result of the action and interaction of various factors, such as the physical characteristics of land, the institutional framework, the infrastructural facilities, the structure of other available resources (capital, labour etc.) and the location of the region in relation to other aspects of economic development including those relating to transport as well as industry and trade [ICAR, 1992: 92]. Table 1 shows that in 1999-2000 net area sown in Barak Valley was 233.8 thousand hectares, which amounts to 33.85% of the total reporting area of the valley. The gross cropped area for the same year was 302.2 thousand hectares. The cropping intensity in the valley was thus 12.9%. In case of Assam as a whole the area under various crops was 4087 thousand hectares enjoying 34.83 percent of the total reporting area. The cropping intensity for the state of Assam as a whole in 1997-98 was 14.9%. From Table 2 it can be seen that the percentage of the Gross Cropped Area (GCA) to the reporting area for Barak Valley in 1981-82 was 38. 9%, which increased to about 44 percent in 1997-98 and 1999-2000, while the respective figures for Assam as a whole in 1981-82 was 44.1%, which increased to 50.88% in 1997-98 and then to 52.07% in 1999-2000. Thus the increasing trend in Assam as a whole is absent in case of Barak Valley.

Area, Population and Households in FA and NFA Region

On the basis of the sample data collected it has been

Table 1
Land Utilization in Assam and Barak Valley, 1999-2000

Classification	*Area (in hect.)*		*Percentage of Reporting Area*	
	Assam	*Barak Valley*	*Assam*	*Barak Valley*
(1)	*(2)*	*(3)*	*(4)*	*(5)*
1. Total Geographical area	7843800	692200	-	-
2. Reporting area	7850005	691097	100.0	100.0
i) Forest	1931631	261741	24.61	37.87
ii) Not available for cultivation	2530925	134445	32.24	19.45
(a) Non-agricultural uses	1069891	74671	13.63	10.8
(b) Barren and unculturable	1461034	59774	18.61	8.65
iii) Other uncultivated land (excluding fallow land)	477368	43617	6.08	6.31
(a) Permanent pastures and other grazing land	162968	6242	2.08	0.90
(b) Miscellaneous tree crops and groves	234206	32963	2.98	4.77
(c) Culturable waste land	80194	4412	1.02	0.64
iv) Fallow land	175620	17441	2.24	2.52
(a) Fallow land other than current fallows	65219	9081	0.83	1.31
(b) Current fallows	110401	8360	1.41	1.21
v) Net area sown	2734461	233853	34.83	33.85
vi) Gross cropped area	4087341	302272	-	-
Area Sown More Than Once	1352880	68419	-	-
Cropping Intensity	149	129	-	-

observed that out of the total area in flood affected region of the Barak Valley the Cachar district occupies 36.74%, Karimganj district 38.14% and Hailakandi district 25.12%. On the other hand, out of total non-flood affected areas Cachar district possesses highest area i.e 49.91%, Karimganj district 39.16% and Hailakandi district 10.93%. Land area-wise Cachar is the largest district in the valley with 45.35% while Karimganjt occupies 38.81% and Hailakandi 15.84%. The flood affected area in the valley is 37.2% and the rest 62.8% non-flood affected area. Out of total households in the FA areas, 36.06% are in Cachar, 40.96% in Karimganj and 22.98% in Hailakandi. In non-flood areas, 56.66% households are in Cachar district, 28.77% in Karimganj and 14.57% in Hailakandi. As a whole, Cachar district occupies 47.8% of households of the Barak Valley, Karimganj district 34% and Hailkandi 18.2%. In Barak Valley as a whole, 43% of the households are in flood affected and the rest 57% are in non-flood affected areas.

Table 2

Gross Cropped Area and its Percentage Share of Reporting Area in Barak Valley and Assam, 1981-82 and 1997-98

Year	*Barak Valley*		*Assam*	
	G.C.A (hect)	*Percentage of the Reporting Area*	*Area G.C.A. (hect.)*	*Percentage of the Reporting Area*
1981-82	268943	38.9	3460082	44.1
1997-98	299191	43.29	3926143	50.88
1999-2000	302272	43.74	4087341	52.07

Sources : 1) Economic Survey 1989-90, Directorate of Economics and Statistics, Govt. of Assam.

2) Statistical Hand Book, Directorate of Economics and Statistics, Govt. of Assam, 1980, 1981, 2000, 2005.

3) Economic Survey, Assam 1996-97, Directorate of Economics and Statistics. Govt. of Assam.

In respect of the distribution of population, the Cachar district accommodates 34.77% of total population in flood affected areas of the Barak Valley, while Karimganj and Hailakandi districts accommodate 41.83 % and 23.40% respectively. Similarly, so far as non-flood affected areas are concerned. 55.34% population resides in the Cachar district, 29.83% in the Karimganj district and 14.83% in the Hailakandi district. As a whole, the Cachar district accommodates 46.3% of total population in the Barak Valley, Karimganj district 35.1% and Hailakandi district 18.6%. So far as the entire Barak Valley is concerned, 43.8% of the total population resides in the flood affected areas and 35.62% population in the non-flood affected areas. Thus around 45% of total households and of population are directly affected by floods in every year. In fact flood has become a natural visitor in every year in the Barak Valley. Though net area affected by floods has come to be about 35%, the loss in crops, livestock and property of the inhabitants of the three districts are enormous. Hailakandi district is the most affected in terms of Area, Population and Households in FA areas. The corresponding figures are 48.3% area, 54.4% households and 55.2% population in FA areas out of its total area, population and households. For detailed statistics and analysis of area, households and population in FA and NFA areas in three districts refer Tables 3 and 4.

Demographic Features

Population of the Barak Valley region as per 1961 census was 13.78 lakh, 17.43 lakh in 1971, 24.91 lakh in 1991 and 29.89 lakh in 2001. This demonstrates some features, which are very much unlike those of the country as a whole, if evaluated in historical perspective and quantifying the validity on the basis of positive data. Between the two regions, the Brahmaputra and the Barak, the former has been generally having a higher growth rate of population. For

Table 3

Area, Household and Population in FA and NFA Areas of Barak Valley

District	Number of villages			Area (in Hectare)			Household			Population		
	(1) FA	(2) NFA	(3) Total	(4) FA	(5) NFA	(6) Total	(7) FA	(8) NFA	(9) Total	(10) FA	(11) NFA	(12) Total
Cachar	269	786	1055	59404	152829	212233	76101	158225	234326	409304	834230	1243534
	(11.40)	(33.40)	(44.80)	(12.70)	(32.70)	(45.40)	(15.50)	(32.30)	(47.80)	(15.20)	(31.10)	(46.30)
Karimganj	447	522	969	61677	119938	181615	86441	80353	166794	492403	449685	942088
	(19.00)	(22.20)	(41.20)	(13.20)	(25.60)	(38.80)	(17.60)	(16.40)	(34.00)	(18.30)	(16.80)	(35.10
Hailakandi	160	171	331	40615	33482	74097	48508	40688	89196	275428	223359	498787
	(6.80)	(7.20)	(14.10)	(8.70)	(7.10)	(15.80)	(9.90)	(8.30)	(18.20)	(10.30)	(8.30)	(18.60)
Barak Valley	876	1479	2355	161696	306249	467945	211050	279266	490316	1177135	1507274	2684409
	(37.20)	(62.80)	(100.00)	(34.60)	(65.40)	(100.00)	(43.00)	(57.00)	(100.00)	(43.80)	(56.20)	(100.00)

*Figures in the parentheses refer to percentages of respective totals.

Table 4
Dirtrict-wise Area, Household and Population in FA and NFA Areas

District	*Number of villages*			*Area (in Hectare)*			*Household*			*Population*		
	(1) FA	*(2) NFA*	*(3) Total*	*(4) FA*	*(5) NFA*	*(6) Total*	*(7) FA*	*(8) NFA*	*(9) Total*	*(10) FA*	*(11) NFA*	*(12) Total*
Cachar	269	786	1055	59404	152829	212233	76101	158225	234326	409304	834230	1243534
	(25.5)	(74.5)	(100.0)	(28.0)	(72.0)	(100.0)	(32.5)	(67.5)	(100.0)	(32.9)	(67.1)	(100.0)
Karimganj	447	522	969	61677	119938	181615	86441	80353	166794	492403	449685	942088
	(46.1)	(53.9)	(100.0)	(34.0)	(66.0)	(100.0)	(51.8)	(48.2)	(100.0)	(52.3)	(47.7)	(100.0)
Hailakandi	160	171	331	40615	33482	74097	48508	40688	89196	275428	223359	498787
	(48.3)	(51.7)	(100.0)	(54.8)	(45.2)	(100.0)	(54.4)	(45.6)	(100.0)	(55.2)	(44.8)	(100.0)

*Figures in the parentheses refer to percentages of respective totals.

instance, during the period from 1901 to 1941 (the year of last census before independence) the population of Barak Valley went up by 36.7%. During the same period the Brahmaputra Valley registered the population growth of 77.7%. This is in conformity with the reality that in the pre-independence era the immigration had been much more pronounced in case of the Brahmaputra Valley than in the Barak Valley. However, between 1941 and 1951, the Barak Valley region experienced a higher growth of population (23.8 percent) than the Brahmaputra Valley (19.93 percent). The apparent explanation for this higher growth rate of Barak Valley is that the valley had to bear the burnt of partition related migration of refugees from East Bengal (now Bangladesh) to a greater extent than the Brahmaputra Valley (Dass, 1980). In the post-independence period the state of Assam has been having a higher growth rate of population than India as a whole. However, for the first time in 2001 census the decadal growth rate (1991-2001) in Assam appears to be less than that of India as a whole. Within the state of Assam, the Barak Valley region has been having a relatively lower growth rate all along (Table 5).

Rural-Urban Compositions

The population of the Barak Valley region is still predominantly a rural one. According to 2001 census, the urban population in the Barak valley region is only around 10%, which is much less than that of the country or even Assam as a whole. In fact, the process of urbanization has been much slower in Assam in general and that of the Barak Valley region in particular. Although the census report lists Karimganj (Class III), Badarpur (Class IV), Silchar (Class I), Lakhipur (Class IV), Hailakandi (Class III) and Lala (Class V) as major towns of the valley, only Silchar is considered as a class I town with a population of more than one lakh. The predominance of rural population also indicates the greater dependence of people here on agriculture and related activities than those of other parts of the country.

Table 5
Decadal Growth Rate of Population in Barak Valley, Assam and All India *(in percentage)*

Year *(1)*	*Barak Valley* *(2)*	*Assam* *(3)*	*India* *(4)*
1901-11	13.80	6.90	5.7
1911-21	5.60	20.5	0.3
1921-31	6.30	19.9	11.0
1931-41	11.00	20.4	14.2
1941-51	23.80	19.9	13.3
1951-61	24.30	34.9	21.6
1961-71	24.20	34.9	24.1
1971-91	45.20	53.3	48.5
1991-2001	19.97	18.85	21.34

Source: Census Report, Govt. of India in various years.
* No census held in Assam in 1981. As such instead of 1971-81, the period 1971-91 is computed.

Workforce and Its Occupational Pattern

The classification of the population of the Barak Valley into the workers, marginal workers and non-workers shows that there is no difference in the working population to total population both in case of rural and urban in the region. As per 2001 census, 89.2% of total population lives in rural areas and only 10.2% in urban areas. Of the total male population in rural area, 48.4% are workers and the figure for urban areas stood at 48%. Similarly, about same proportion (7 percent) are female workers both in the rural and urban areas. About 13.55% of total female population is workers and 49% of male population is non-workers. This also clearly speaks the grim state of unemployment scenario in the region. The 2001 census returned about 70.0% of the total working force of the Barak Valley region as engaged in primary sector (agriculture and allied activities), 7.0% in secondary sector and 23.0% in tertiary sector. Thus agriculture and allied activities are still the predominant source of livelihood of the population. Only about 7% of the workforce being engaged in the secondary sector is an indicative of the underdevelopment of the region in manufacturing and other secondary activities. A detailed comparative analysis of work force participation—main and marginal workers—is shown in Table 6.

Table 6

Workforce Participation, Distribution of Main and Marginal Workers

(as per 2001 census)

	Workforce participation rate			*Main Workers*			*Marginal Workers*		
	Rural	*Urban*	*Total*	*Male*	*Female*	*Total*	*Male*	*Female*	*Total*
(1)	*(2)*	*(3)*	*(4)*	*(5)*	*(6)*	*(7)*	*(8)*	*(9)*	*(10)*
Cachar	32.33	31.22	32.18	41.85	7.52	25.17	15.90	44.56	21.77
Karim-ganj	30.33	31.48	30.41	40.40	5.30	23.36	15.70	55.30	23.20
Hailak-andi	33.66	29.78	33.33	41.55	8.10	25.41	15.18	51.04	23.77
Barak Valley	32.11	30.83	31.97	41.27	6.97	24.65	15.59	50.30	22.91
Assam	36.4	32.0	35.9	42.35	9.68	26.59	7.58	11.12	9.29
All India	42.0	32.2	39.3	45.35	14.68	30.55	6.59	10.99	8.71

Source: Assam Human Development Report, 2003.

Economy - The Sectoral Composition

In 2000-01, Barak Valley accommodated 11.71% of the population of Assam while its share in Gross State Domestic Product (GSDP) was only 9.33% .Obviously, the district per capita income is below the state average, which in turn is well below the national average. As per statistics of 2000-01, the per capita income (i.e. net district domestic per capita) in Barak Valley was Rs.5223 at 1993-94 prices. The figure comes up to only about 84% of the corresponding per capita income figure for Assam. The details of District Domestic Product (DDP) and its composition in 2000-01 are shown in Table 7.

The sectoral composition of the economy confirms a strong agricultural bias. Agriculture contributed 33.1% Gross District Domestic Product (GDDP) and the primary sector contributed 38.3% of the same. The tertiary sector comprising of the various services constitutes more than one third of the GDDP. The manufacturing sector's share is less than 8.7% of the secondary sector as a whole and it accounts for 9.49% of the GDDP. The relatively high share of secondary sector (higher than that of the manufacturing sector) is due to a relatively higher contribution of construction activities. The sectoral composition of GDDP of Barak Valley has a strong resemblance to the sectoral composition of Gross State Domestic Products (GSDP) of Assam. For instance, for the state as a whole too agriculture and primary sector still constitutes the mainstay. Moreover, in the secondary sector the construction activities played an important role. Thus as far as sectoral composition of domestic product is concerned, Barak Valley does not show a far greater backwardness than the state as a whole, though compared to the country both Barak Valley and Assam appears to be far more backward. However, relative backwardness of Barak Valley compared to Assam can be inferred if the sectoral composition of domestic product is matched with the occupational pattern of the work force of the valley. It has been found that the employment of the secondary activities is only meagre 6.26% of the work force of the Barak Valley in spite of this sector contributing nearly 17% of the Gross District Domestic Product. Agriculture continues to employ about 70% of the work force though the sector's contribution to Gross District Domestic Product is only about 38%. Thus sectoral composition of the economy as revealed by the composition of the GDDP is not matched by the occupational pattern of the work force. It may be noted that for Assam as well as for India, the shift in the occupational pattern of work force from agricultural to non-agricultural occupations has been less than proportionate to the corresponding change in sectoral composition of domestic product. However, for Barak Valley the occupational pattern seems to have shifted far too slowly. The implication of the observation is that the gains from the non-agricultural sector are concentrated in the hands of a few leaving the vast majority of the population still dependent on agriculture.

Manufacturing Industries

The region's backwardness in manufacturing activities can be understood from the fact that the household and non-household industrial units engaged in manufacturing, processing, servicing and repairing employ only 6.0% of the total work force (as per 2001 census). The most of the industrial units outside the household sector are either agriculture or forest based. In private factory sectors, the notable industrial units of the region are tea manufacturing units, plywood and saw mills. The notable public sector units set up in the region since independence are Panchgram Paper Mill under Hindustan Paper Corporation in Hailakandi district and Cachar Sugar Mill in Karimganj district. Of the two, the Sugar Mill was closed down within few years of commissioning. The valley has a long tradition in handicrafts, weaving, carpentry, boat-making and also in the manufacture of a variety of consumer goods. The availability of cheap

Table 7
Gross District Domestic Product (GDP) of Barak Valley: 2000-01

	GDDP at 1993-94 prices (Rs. In Lakhs)	*NDDP at 1993-94 prices (Rs. in Lakhs)*	*Contribution of sectors to GDDP (%)*	*Contribution of Barak Valley to GSDP of Assam (%)*
(1)	*(3)*	*(5)*	*(6)*	*(7)*
A. Primary (Total)	63617	61136	38.30	8.86
1. Agriculture	54985	53334	33.10	10.48
2. Forest & Logging	2679	2642	1.61	8.08
3. Fishing	5510	4832	3.31	4.59
4. Mining & Quarry	443	328	0.27	0.36
B. Secondary (Total)	28186	34047	16.97	10.20
1. Manufacturing	14506	11758	8.73	9.49
a. Registered	13691	11114	8.24	11.48
b. Un-registered	816	643	0.49	2.29
2. Construction	12157	11182	7.32	12.32
3. Electricity & Water-Supply	1523	-650	0.92	6.44
C. Tertiary (Total)	74310	66158	44.73	9.44
1. Railways	1045	169	0.63	6.57
2. Storage	-5	-5	-0.003	-3.59
3. Transport by other means	4325	1608	2.60	11.57
4. Communication	3746	2676	2.26	14.28
5. Trade, Hotel & Restaurant	25465	24461	15.33	10.73

(Contd...)

(1)	*(2)*	*(3)*	*(4)*	*(5)*
6. Banking & Insurance	10442	10114	6.29	12.47
7. Real Estate, Owner-ship of Dwellings & Business Services	2016	1609	1.21	3.22
8. Public Administration	10024	8862	6.03	7.61
9. Other Services	17247	16664	10.38	8.95
Total	166113	161341	100.0	9.32
Per Capita Income (in Rs.)	5377	5223		

Source: Directorate of Economics & Statistics Govt. of Assam.

skilled and semi skilled labour and the prevalence of a favourable law and order situation, offer a somewhat favourable environment for growth of tiny and small industrial units. However, the geographical isolation of the valley, the fragility of the existing infrastructure particularly power, transport and communication has always been acting as a drag on the economic development of the region [Mazumdar et al., 1998: 5] As such, the prospects of economic development of the region depend to a great extent on utilization and exploitation of growth potential of agriculture in the region and development of agro-processing industries.

Agriculture: The Cropping Pattern

The cropping pattern is indicated by the shares of different crops in the gross cropped area. This may undergo changes over time in response to changes in economic, technological or other environmental factors. A striking feature of cropping pattern of Barak Valley region is the predominance of paddy, which accounts for more than 90 percent of above defined total cropped area during the period 1971-72 to 1999-00. The share of paddy during these periods fluctuated between 91.64% and 93.39% without showing any distinct trend. Other notable crops cultivated in the valley are pulses, rape and mustard, potato and sugarcane (Table 8).

Table 8
Percentage Share of Principal Crops in Total Cropped* Area in the Barak Valley Region during the Period 1971-72 to 2004 05

CROPS	*1971-72*	*1979-80*	*1984-85*	*1990-91*	*1994-95*	*2001-02*	*2004-05*
(1)	*(2)*	*(4)*	*(5)*	*(7)*	*(11)*	*(15)*	*(16)*
Autumn Rice	16.55	19.36	13.85	16.49	11.20	8.47	8.95
Winter Rice	71.32	68.40	72.52	70.95	76.30	73.48	72.11
Summer Rice	5.52	4.94	5.27	5.64	5.20	5.59	7.18
Total Rice	93.39	92.70	91.64	93.08	92.70	87.54	88.23
Other Cereals	0.15	0.20	0.12	0.11	0.06	0.11	0.11
Total Cereals	93.54	92.90	91.76	93.18	92.76	87.64	88.34
Pulses	0.80	0.93	1.41	1.32	2.36	2.38	5.69
Total Foodgrains	94.34	93.83	93.17	94.50	95.12	90.02	94.03
Rape and Mustard	0.66	0.80	1.78	1.77	0.93	1.04	1.23
Other Oilseeds	0.16	0.20	.0.20	0.15	0.27	0.13	0.16
Total Oilseeds	0.82	1.00	1.98	1.92	1.20	1.17	1.39
Total Fibre	0.50	0.35	0.29	0.17	0.15	0.14	0.17
Potato	1.13	1.11	1.74	1.42	1.63	1.86	1.96
Sugarcane	2.12	2.76	1.86	1.24	1.17	0.53	0.48
Others	1.09	0.96	0.96	0.75	0.73	6.27	1.97
Total	100.0	100.0	100.0	100.0	100.0	100.0	100.0

* Excludes area under plantation and other tree crops like banana, papya etc.Sources : 1) Estimates of Area, Production and Average Yield of Principal Crops in Assam from 1951-52 to 1987-88, Directorate of Economics and Statistics, Govt. of Assam.

2) Basic Agricultural Statistics, Directorate of Agriculture, Govt. of Assam, Various issues (1989-90 to 2004-05).

Paddy, which dominates the cropping pattern, does not however constitute just one crop. In fact on the basis of the harvesting seasons, distinctions are drawn between three different paddy crops, namely, autumn paddy, winter paddy and summer paddy. In the context of the Barak Valley region, winter paddy consists mainly of transplanted 'Sali' crop. It is transplanted from nursery beds in July and August and harvested in November and December. A lesser known winter paddy-crop, locally known as *'birain'*, is cultivated in broadcast method in low-lying areas. *'Birain'* paddy has the special ability to grow taller with the rise in the water level and can withstand floods to a great extent. It is sown during March and April and harvested in July- August. Of the three categories of paddy, winter paddy is most important in terms of shares in area under paddy. For instance, in 1999-00, winter paddy comprised 85.37% of total area under paddy in the region, while autumn paddy comprised 9.49% and summer paddy only 5.14% of total paddy acreage of the region. It is observed that, in 1971-72 winter paddy constituted 76.37%, while autumn paddy and summer paddy constituted 17.72% and 5.91% respectively of total area under paddy. Thus it is clear that winter paddy has increased its share of total area under paddy over the years. Correspondingly the share of autumn paddy has declined from 17.72% in 1971-72 to 9.49% in 1999-00 (Table 9). The share of summer paddy remains more or less unchanged at 5%. It is interesting to note that these trends are somewhat in contrast with the broad trends observed in the state as a whole. For the state as a whole there has been a significant increase in the share of summer paddy during the reference period of our study with a result of decline in the share of winter paddy. Although traditionally the Barak Valley region had a much higher share of summer paddy in total paddy area than for the state as a whole, there has been no further improvement in the share in the Barak Valley region like in the other parts in the state. Comparatively higher traditional share of summer paddy in the region than in the state as a whole can be attributed to existence to a large extent of low-lying natural depression and waterlogged areas suitable for cultivation of '*boro*' (popularly known in the region by the farmers as '*borua*') paddy. However, the absence of the increase in the share of summer paddy area in the region seems to be due to lack of progress of the region in respect of irrigation infrastructure [Roy and Bezbaruah, 2002].

Principal Crops

In the trend analysis in agricultural production of the Barak Valley region for the present study seven crops grown in the region have been included. The selection of these seven crops have been made on the criterion that each of them commanded at least 1% or more of total cropped area for at least one year during our reference period 1979-80 to 1999-00. The crops included are autumn rice, winter rice, summer rice, pulses, sugarcane, rape and mustard and potato. The estimated annual compound growth rates of different crops included in the study have been presented in Table 10

Table 9

Distribution of Area Under Different Categories of Paddy to the Total Area under Paddy in the Barak Valley Region and Assam (percentage share)

	Barak Valley (percentage to the total area)			*Assam (percentage to the total area)*		
Year	*Autumn*	*Winter*	*Summer*	*Autumn*	*Winter*	*Summer*
(1)	*(2)*	*(3)*	*(4)*	*(5)*	*(6)*	*(7)*
1971-72	17.72	76.37	5.91	26.77	71.62	1.61
1974-75	22.44	71.56	6.00	27.38	71.10	1.52
1979-80	20.90	73.80	5.30	25.42	72.84	1.74
1984-85	15.11	79.14	575	26.11	71.85	2.04
1989-90	14.97	78.30	6.73	25.44	71.42	3.14
1992-93	13.31	80.67	6.02	24.82	69.40	5.78
1994-95	12.08	82.31	5.61	26.00	68.70	5.30
1999-00	9.49	85.37	5.14	24.24	66.63	9.13
2004-05	10.14	81.72	8.14			

Sources: 1) Estimates of Area, Production and Average Yield of Principal Crops in Assam, from 1951-52 to 1987-88, Department of Economics and Statistics ,Govt of Assam.

2) Basic Agricultural Statistics, Directorate of Agriculture, Govt. of Assam

Table 10
Annual Compound Growth Rates of Different Crops in the Barak Valley Region and Assam between 1979-80 to 1999-00, Growth rate (in percentage)

Crops	Area	Production	Yield
(1)	(2)	(3)	(4)
1. Autumn Rice	-2.85 (0.34)	-1.71 (1.68)	1.17 (1.34)
2. Winter Rice	0.75 (0.81)	3.52 (2.35)	2.75 (1.54)
3. Summer Rice	1.51 (7.61)	1.34 (10.39)	-0.17 (2.77)
Total Rice	0.15 (0.89)	2.46 (2.51)	2.30 (1.62)
4. Pulses	5.99 (1.93)	7.58 (3.16)	1.50 (1.23)
5. Sugarcane	-6.22 (-0.85)	-4.86 (0.14)	1.46 (1.00)
6. Potato	2.95 (5.22)	2.64 (7.48)	-0.30 (2.26)
7. Rape and Mustard	8.07 (4.10)	8.24 (5.52)	0.15 (1.42)

Note: Figures in the parentheses represents respective growth rate for the state as a whole.

A striking feature of the growth rates presented in Table 4 is the rapid growth of production of the two crops, namely, rape and mustard and pulses. The high growth of production of the former is almost entirely due to growth in area, yield growth being practically negligible. In case of pulses too, area expansion is the main source of growth of production, though a growth of yield at 1.5% per annum also contributed significantly. The production of rice (three crops taken together) and potato also registered impressive annual compound growth rates of 2.46% and 2.64% respectively. The growth in rice production was made possible by growth in yield at the rate of 2.3%, as area under rice did not expand significantly. In case of potato, however, the situation is reverse. There was a mild declining trend in the yield of potato, but there was a significant growth of the area at about 3% per annum. In contrast to the crops already mentioned, there was a high negative growth of production of sugarcane in spite of a positive growth in its yield at the rate of 1.46% per annum. There was a massive shift of area from sugarcane to other crops during the reference period, the growth rate of area under sugarcane being –6.22% per annum. As already mentioned in the previous section the sickness and closure of the Cachar Sugar Mill seems to have had a telling impact on the area allocation to sugarcane by the farmers.

These broad trends in growth of the crops are in conformity with those observed in the state level. However, as one takes a closure look on the pattern of growth of rice crops, a contrasting picture appears for the Barak Valley region and the state as a whole. The growth and production of rice in Barak Valley mostly occurred in the winter rice variety. The growth rate of production of winter rice at the rate of 3.52% per annum in the Barak Valley was substantially higher than the rate of 2.32% per annum for the state as a whole. The area growth rate was more or less at par and below 1% for both the Barak Valley and Assam as a whole. The yield growth rate was much higher for Barak Valley (that is 2.57 percent per annum). An important feature of the growth pattern of rice in Assam during these periods is the high growth rate of production of summer rice resulting from fast expansion of area and an impressive growth of yield of this crop in the Brahmaputra Valley of the state. A similar pattern is completely absent in the Barak Valley region. Here not only was the growth of production of summer rice much slower, but the growth came entirely due to area expansion as there was a mild declining trend in the yield of this crop. In case of autumn rice there was some growth in yield for both Assam and the Barak Valley region. But unlike the state as a whole, in Barak Valley there was a significant negative growth in the production of autumn rice. The decline in the production in spite of positive growth of yield is accounted for by a significant negative growth of area at the rate of –2.85% per annum.

Irrigation Infrastructure

Irrigation perhaps is the most crucial component of agricultural infrastructure. The HYV fertilizer technology based strategy for agricultural growth through increase in crop yield and cropping intensity depends critically on irrigation. To increase in cropping intensity and for reduction of instability in crop production through greater dependence on flood free crops, the development of irrigation infrastructure in the region in an imperative. A detailed scenario of availability of irrigation is shown in Table 11. Table 12 shows irrigation potential available and irrigation potential utilized in the Barak Valley region during the period 1990-91 to 2000-01. A close look into the figures in the table shows a grim state of irrigation infrastructure in the valley. Though irrigation potential available improved from 1.91% of estimated gross cropped area in 1990-91 to 4.09% in 2000-01, the ratio still remains extremely low to have any significant impact on the agriculture of the valley. Moreover, the utilization rate of this meagre irrigation potential created has fallen steadily over the years and utilization rate was as low as 1.96% in 2000-01.

Table 11
Irrigated and Unirrigated Areas of Barak Valley, 2004-05

	Irrigated area	*Un-irrigated area*	*Net Area Sown*
1	*2*	*3*	*4*
Karimganj	1945.47 (2.3)	82848 (97.7)	84793.47 (100.0)
Cachar	1809.1 (1.5)	121692.39 (98.6)	123451.5 (100.0)
Hailakandi	126.26 (0.3)	47072.49 (99.7)	47198.75 (100.0)
Barak Valley	3880.83 (1.5)	251612.88 (98.5)	255493.7 (100.0)

Table 12
Irrigation Potential Available and Irrigation Potential Utilized in the Barak Valley Region during the period 1990-91 to 2000-01

Year	*Irrigation potential available during the year (in hectares)*	*Irrigation potential utilized during the year (in hectares)*	*Estimated utilization ratio (as percentage)*
(1)	*(2)*	*(3)*	*(4)*
1990-91	5610 (1.91)	5066 (1.73)	90.30
1991-92	7274 (2.51)	4421 (1.52)	60.78
1992-93	6726 (2.28)	3533 (1.20)	52.53
1993-94	7064 (2.42)	698 (0.24)	9.88
1994-95	5831 (2.06)	1248 (0.44)	21.40
1995-96	8024 (2.83)	800 (0.28)	9.97
1996-97	6299 (2.23)	283 (0.10)	4.49
2000-01	12380 (4.09)	243.07 (0.08)	1.96

Sources: Office of the Chief Engineer, Irrigation, Govt. of Assam.

Note: Figures in the parentheses are percentages of estimated total cropped area of respective years.

Trans-border Trade

The Barak Valley region is endowed with one hundred kilometers border with Bangladesh. A large volume of border trade of the Northeast with Bangladesh takes place through the two border transit points of the Karimganj district, namely, Sutarkandi (Land Custom System) and Karimganj Steamerghat Land Custom System (riverine route). Karimganj is the only border trade point in Assam between India and Bangladesh. It is to be noted that before 1965 Karimganj border had four Land Custom Stations (LCS), namely, the Mohishasan LCS (rail route), Sutarkandi LCS (land route), Karimganj Steamerghat LCS (riverine) and Karimganj Ferryghat LCS (land route). After 1965, the riverine route via East Pakistan was closed, and Karimganj Steamerghat LCS was amalgamated with Karimganj Ferryghat LCS.

With the SAARC initiative the volume of trade with Bangladesh has increased since 90's. Due to the logistic advantage, traditionally trade with Bangladesh was mainly informal (Dutta and Roy, 2000). Things have changed during the past couple of years due to the liberalization in trade restrictions amongst the SAARC countries. As a result, volume of trade and balance of payments has improved. In keeping with the overall Indo-Bangladesh trade trend, the Karimganj LCS also confirms favourable balance of trade and improving terms of trade. Data collected from the Customs Department, Karimganj, about export and import is depicted in Table 13.

The major export items were coal, non-basmati rice, fresh ginger, dry chillies, citrus (local variety called *Hatkada*), fresh vegetables, jackfruit, banana, grapes, pineapple, fresh orange, betel leaves and nuts, mango, peas, raw cotton, soyabean, onion, garlic, turmeric, Rohu fish (Andhra variety), grain sorghum, tube light, readymade dress materials, parts and accessories of tractors and motor vehicles, umbrella parts, soap, aluminum and brass products, poultry feed etc., while imports consisted mainly of Hilsa fish, chita, molasses, dhakai sari, fruit juice, cold drinks etc. While coal comes from neighbouring state of Meghalaya, fresh vegetables and fresh fruits originated from Mizoram and Manipur and other industrial products come from northern states. It is to be noted that the demand for agricultural commodities like fresh vegetables and fruits have increased manifold over the years. The entire north eastern region can take the advantage of these growing demands of agricultural commodities in Bangladesh. The Northeast possesses tremendous potentials of production of exportable items like ginger, citrus, pineapple, orange etc. but these potentials are yet to be explored fully.

There is however some peculiar problems of border trade between India and Bangladesh that have been noticed over the years in the trade points. Informal trade and under-invoicing are major irritants which hamper growth of trade notwithstanding trade liberalization measures.

Table 13
Value of Export Import Trade with Bangladesh (Karimganj Border Point)

Rs.Crores

Year	*Sutarkandi LCS*		*Steamerghat LCS*				*Total Bal. of Payment*
	Export	*Import*	*Export*	*Import*	*Export*	*Import*	
1996-97	21.67	4.70	4.29	0.0008	25.96	4.7008	+21.26
1997-98	33.76	4.56	9.48	nil	42.24	4.56	+38.68
1998-99	28.63	5.30	22.62	0.05	51.25	5.35	+45.9
1999-00	35.38	3.93	2.12	nil	37.5	3.93	+33.57
2000-01	18.38	1.68	2.63	0.01	22.01	1.69	+19.32
2001-02	13.79	00.73	3.91	00.33	17.70	1.06	+16.64
2002-03	22.87	00.08	6.00	00.13	28.87	00.21	+28.66
2003-04	48.14	1.63	2.03	0 .004	50.77	1.637	+49.133
2004-05	21.34	6.61	3.64	0.07	24.98	7.67	+17.31

Source: Department of Customs, Karimganj.

Best-case Scenario for Barak Valley

Based on the analysis from the study presented here the best scenario for the valley may be visualized as follows:

Agriculture - Productivity revolution in agriculture of the region with diversification to commercial crops, agri-business, processing industries and agro-exports will generate abundant employment in the rural sector. The transformation of rural sector will stimulate the process of urbanization in the region leading to expansion of the consumer goods and service sector. The export-import market will boost diversification in agricultural sector.

Industry - The industrialization is the key to economic development and an answer to the mounting unemployment problem of the valley. The local availability of raw-materials, long tradition in handicrafts, weaving, carpentry, boat-making and also in the manufacture of a variety of consumer goods, availability of cheap skilled and semi skilled labour, and the prevalence of a favourable law and order situation in the valley offer favourable environment for growth of small and medium industrial units, particularly the forest- and agro-based industries. However, the development of communication network and the improvement of the power sector should get priority to facilitate industrial growth of the valley. The appropriate infrastructure is also likely to attract investment from outside.

Plantation- The revival of tea plantation will boost tea trading of Barak Valley with other countries of the world. This will enhance employment opportunities. At the same time rubber plantation will add a new dimension in the economy of the valley.

Trans-border trade – Trans-border trade with Bangladesh will successfully integrate Barak Valley with world economy. The Sutarkandi has been declared as Special Economic Zone (SEZ). The Ministry of Commerce, Government of India has also undertaken development of infrastructure in Sutarkandi. This is in line with the SAFTA agreement amongst the SAARC countries.

Infrastructure Development & Power - Despite poor power position in the state of Assam a slow but steady progress in power development has been noticed over the past few years. In the Barak Valley region itself, two gas based power projects (one at Adamtila in Karimganj district and another at Banskandi in Cachar district) have already started generating power. There is also a proposal for construction of the Barak Dam Hydro-Electric Power Project at Tipaimukh on Cachar-Manipur border, which is likely to be undertaken by the Government of India immediately. If these projects be materialized, the existing power crisis in the region will be solved to a great extent and the additional amount can be supplied to other parts of the state.

Transport - The most tragic feature is that the entire Barak Valley has remained very poor in this economic infrastructure. A detailed account of transport and communication facilities available in the Barak Valley region is given below:

i) *Rail and Road Network* - The Barak Valley is yet to get a Broad Gauge railway connectivity. However, construction works of Broad Gauge railway line between Lunding and Silchar is in progress. The percentage of surfaced road to total roads in the state of Assam is 31.02% and in the Barak Valley it is approximately 30.74% in 2002. There is possibility of a short-cut Asian Highway link through Sylhet-Sutarkandi-Karimganj-Imphal-Tamu. Since

Bangladesh has a unique highway from Seola (opposite side of Sutarkandi) which connects Dhaka via Sylhet, the connectivity between Sutarkandi and Seola can provide a link between two countries and boost the border trade. The region can also be connected to the rest of India via Bangladesh by both roadways and waterways. However, the improvement of internal roads within the valley and the development of the highways connecting the valley with the neighbouring states and other parts of the country should be an important strategy.

ii) *Water Transport* - Before the partition, water routes used to be an important mode of transport of goods between Barak Valley and the rest of the world through the rivers in East Bengal. These waterways continued to be active till the Indo-Pak War of 1965. But the volume of transport through these waterways has since declined considerably to a negligible level. Even today, some transportation by steamer takes place between Karimganj in Barak Valley and the Calcutta Port through the traditional waterways. But this takes place only in the summer months. The water transport system is currently operated by the Central Inland Water Transport Corporation (CIWTC). The region has good potential for revival of inland water transport routes.

iii) *Air Transport* - Of the six civil airports in Assam, the Barak Valley region shares one, at Kumbirgram, 25 km from Silchar town. The valley is served by the Indian Airlines, which provides regular air service between Kolkata and Silchar and other capital towns of the Northeast. The Air Deccan has now introduced daily air connectivity between Kolkata and Silchar.

The process of upgradation of the Silchar airport is in process.

Human Resources - Inequalities between different age groups, the sexes, income groups, communities and regions are showing a drastically declining trend globally. The education is playing a decisive role in moving forward together towards a more secure and prosperous future. The Barak Valley has a long tradition in education with prestigious institutions. The challenge is to modernize the institutions and the courses to make education relevant to social change and social development.

Resume

The general features of the economy of the Barak Valley region shows that the potentials are altogether not absent in the valley for new entrepreneurial ventures. The valley has a long tradition in handicrafts, weaving, carpentry, boat-making and also in the manufacture of a variety of consumer goods. The availability of raw-materials and of cheap skilled and semi skilled labour and the prevalence of a favourable law and order situation, offer a friendly environment for growth of small and medium industrial units. However, the prospects of economic development of the Barak Valley region depend to a great extent on utilization and exploitation of growth potential of agriculture in the region. The geographical isolation of the valley, the fragility of the existing infrastructure (particularly power, transport and communication) which are acting as a drag on the economic development of the region are negotiable, provided the political will is sympathetic to the needs and aspirations of the people of the Barak Valley.

REFERENCES

Bezbaruah, M.P. (1997), "Regional and Seasonal Variations in the Trends in Rice Production in Assam from 1974-75 to 1994-95" (summary), *Indian Journal of Agricultural Economics*, Vol. 52, No.3, July-September.

Bhattacharjee, J. B. (1977), *Cachar Under British Rule in North East India*, Radiant Publishers, New Delhi.

Bhattacharjee, J B (ed) (1989), *Sequence of Development in North East India*, Omsons Publication, New Delhi.

Dandekar, V.M. (1980), "Introduction", *Indian Journal of Agricultural Economics*, Vol.35, No.11, April - June.

Das, M.M. (1984), *Peasant Agriculture in Assam*, Inter India, New Delhi.

Dass, S. (1980), *Spotlight on Assam*, Premier Book Service, Chanderpur, Maharastra.

Dass, Susanta (1999), "Trans-Border Trade via Karimganj, 1947-98", *The North East Twilight*, Silchar, 30th September.

Dutta, P.C and Niranjan Roy (2000), 'The Impact of Cross- Border Human Traffic: A case study of Karimganj district' in Gurudas Das and R.K. Purkayastha (ed), *Border Trade – North East India and Neighbouring Countries*. Akansha Publishing House, New Delhi.

Goswami, Atul (1986), "Land Reforms in Assam : A Critical Review" in Atul Goswami (ed.), *Land Reforms and Peasant Movement*, Omsons Publications, Guwahati.

Indian Council of Agricultural Research (1992), *Hand Book of Agriculture*, New Delhi.

Indian Council of Social Science Research (1980), *A Survey of Research in Economics*, Vol. VI, New Delhi.

Mazumdar, Apurbananda, Prasanta Rn. Acharjee and Joydeep Bhattacherjee (1998), *Statistical Profile of Barak Valley 1998*, North Eastern Centre for Advanced Studies, Silchar.

Rao, V.K.R.V. (1989), "Inaugural Address of Golden Jubilee

Conference of the Indian Society of Agricultural Economics", *Indian Journal of Agricultural Economics*, Vol. 44, No. 40, October-December.

Roy, Niranjan and M P Bezbaruah (2003), *Agricultural Growth and Regional Economic Development*, Mittal Publications, New Delhi.

Vyas, V. S. (1968), "Economic Efficiency on Small Farms of Central Gujarat", *Seminar Series, No. 7*, Indian Society of Agricultural Economics.

6

Economic Development of Barak Valley: Tea Industry as the Leading Sector

—Alok Sen

The erstwhile Cachar district of Assam, which has now been reorganised into three separate districts - Cachar, Karimganj and Hailakandi, is now popularly known as Barak Valley since the area is spread over both banks of river Barak till the foot of the neighbouring hills. Though reorganised for administrative convenience, all the three districts exhibit common features – geographical, economic, cultural and linguistic. Originally this region along with the Sylhet district (now four districts) of present Bangladesh was a common administrative division, known as 'Surma Valley Division'. The economy of that undivided valley was quite balanced and well integrated with the mainstream economy of India. The leading sectors in the economy were agriculture and tea industry.

It is often said that there was a time when economy of the Barak Valley was revolving around the tea industry. The soil and climate of this valley were suitable for tea cultivation and the British took full advantage of those geo-climatic conditions. Tea industry was flourishing and it brought about a complete change in the economy of this valley. Though started 20 years after (1855) Brahmaputra Valley, tea industry of Cachar soon became a leader in this field. During the nineteenth century, Cachar was holding the top position in production among the tea producing districts of Assam. In 1869 Cachar had 24151 acres of land under tea and yielded 4255000 lbs. of finished tea. In the beginning the production in Brahmaputra Valley advanced more rapidly than that of the Surma Valley (Sylhet & Cachar) but since 1876 the balance dramatically reversed [Bhattacharjee, 1977 : 199]. In 1882 the area under cultivation in Cachar had risen to 48873 acres and the productivity to 1,27,21,000 lbs which was the highest production from any district of Assam. The annual growth rates during this period were as high as 15.5% and 23% for the area under tea and production respectively. In 1885, out of the total production of 5,36,17,020 lbs in Assam province, Surma Valley with only two districts, Cachar and Sylhet, yielded 2,09,98,978 lbs, while Brahmaputra Valley which had six districts could produce only 32618042 lbs. Of the tea producing districts in Assam, Cachar topped the list and was followed by Sibsagar, Lakhimpur and Sylhet. In respect of tea gardens also Cachar had the distinction of having highest number of tea gardens (199 nos)., followed by Sibsagar (174), Lakhimpur (141) and Sylhet (133). In 1900 the area under tea cultivation in Cachar was 61,179 acres with an average productivity of about 500 lbs of manufactured tea.

In the sphere of export and price realization, the performance of Cachar tea was remarkable. India was the largest exporter of tea, and 'Cachar tea' was holding a dominant position among the manufacturing regions. In the price front Cachar tea were fetching very good price and was strong enough to resist any price and demand crash. We can quote the remarks of John Smeal, proprietor of several tea gardens in Cachar, to the then Commissioner in 1868, "... our teas have fetched a very high price, averaging last year in London, rupees 125 a maund. Rupees 60 per maund is a common average price in London for Cachar Tea" [Griffith, 1967 : 91]. Despite affect of crisis in 1860s, the whole tea industry of Cachar exhibited a comparatively better performance in partially resisting the crisis and making possible an early revival. The opinion of Edgar made in 1873 is of worth mention here : "To that keen observer, the prospect of tea seemed more hopeful in Cachar than any other district." Edgar's main ground for this belief was that the planters of Cachar "seemed to have made more progress than any other district in utilizing scientific method in cultivation and manufacture for the crude experiments which caused more loss in earlier times" [Griffith, 1967 : 99]. He also considered the climate of Cachar more suitable for tea than that of Assam Valley or the Terai [Griffith, 1967 : 99].

Table-1

District-wise Growth of the Tea Industry in Assam

District	*Acreage under tea (figures in thousand acres)*						
	1883	*1900*	*1920*	*1935*	*1945*	*1950*	*1955*
Sylhet	28	71	91	89	6	-	-
Cachar	50	61	63	53	75	75	73
Goalpara	-	1	1	4	4	4	4
Kamrup	6	4	4	4	5	5	5
Darrang	17	42	57	61	62	64	64
Nowgong	11	13	13	12	12	13	13
Sibsagar	41	78	99	103	104	109	106
Lakhimpur	37	68	61	109	112	116	118
Total	109	337	420	435	373	386	386

Source: *Digest of Tea Statistics, 1971-72*, pp. 4-5.

It reveals from the facts above that the tea industry in this valley has a tremendous potential to progress, but what is happening now is just the reverse. Tea industry of Barak valley, as it is found now, is in a very bad shape. The productivity in Barak Valley is the lowest in Northeast India. About 50 percent of the tea bushes are above the age group of 50 years, re-plantation is about 0.5 percent as against the suggested rate of 2 percent per annum. The matter of greater concern is the existence of 28 sick tea estates out of 102 estates in the valley, constituting 20 percent of the total tea grant area. Moreover, out of the total tea grant area only 39 percent have been utilised so far for growing tea. In terms of productivity (Kg/hec.) the performance of Barak Valley comes out to be very disappointing when it is compared with the other major tea producing regions of India (Table- 2).

Table-2
Average Productivity of Tea in Assam and Other Places in India

(in kg/hectare)

District/ state	*1951*	*1961*	*1971*	*1981*	*1991*	*1995*	*2005*
Darrang	1063	1177	1336	1655	2016	1999	2007
Goalpara	1102	1071	1158	1386	1697	1820	1576
Lakhimpur	1193	1460	1605	1901	1962	1849	1720
Dibrugarh				2048	2135	2195	2164
Cachar	651	647	733	904	1150	1341	1403
Total Assam	966	1123	1227	1503	1717	1779	1743
Terai	759	990	1203	1395	1697	1896	-
Dooars	1711	1222	1359	1581	1732	1755	-

Source: Tea Statistics 1956-57.

NB : Cachar includes Cachar (present), Hailakandi and Karimganj districts; Lakhimpur includes Dibrugarh and Lakhimpur (North) districts.

It is observed in the above table that the progress of Cachar tea during this period is very unimpressive while the other districts like Lakhimpur and Sibsagar making quite significant progress. Cachar had also a decline in acreage during the period from 1920 to 1935 by 10 thousand acres and from 1950 to 1955 by 2 thousand acres. The acreage growth rate of Cachar during the period was 46% whereas that attained by Lakhimpur district was 218%. The overall growth rate of acreage attained by Assam was 104%.

In order to find out the causes of such deterioration of the tea industry in Barak Valley this paper intends to adopt a Stationary Test type of technique. It attempts to single out first the major factors which are pre-requisites for the success of any industry. In economic literature, the buzz words of all the pragmatic development models are: natural endowments, entrepreneurship, capital investment and the infrastructure like communication, electricity, etc. These factors are addressed below :

1) **Natural Endowments**: The natural endowments of this region remained almost the same as in the nineteenth century. There could be some minor environment degradation but that has certainly not brought any significant change in the geo-climatic features of the valley. Cachar, which was once found by the experts as very suitable - with its climate and land - for the cultivation of tea (annexure-1), cannot suddenly turn unfavourable now for the same purpose. So we can logically conclude that Barak Valley till today contains huge growth potential in the form of her natural endowments, which if properly harnessed can certainly help to scale the economy to a new height. But unfortunately this is not happening in this valley. On the contrary, since Independence, the situation is turning from bad to worse. Besides

tea, we have plenty of fertile agricultural land, but our productivity is very low, because we have failed to modernise the technique of cultivation and to diversify our product. We have tremendous potential for developing eco-tourism but till to-day there is no avenue for the local people even to spend few hours for recreation.

2) **Entrepreneurship:** The major role in developing an economy lies with the entrepreneurs. It is true that the entrepreneurs' basic intention may not be the development of a region, but in the process of value addition of their invested capital - by the operation of various forward and backward linkages - the economy of the region is sure to get activated and thus, to move forward. The British entrepreneurs ventured into the tea sector of this region in the nineteenth century mainly for earning profit, but there cannot be any denying of the fact that the whole economy moved forward along with tea industry. But since the first quarter of the twentieth century some withdrawal of interest among the British investors was observed and the British capital were gradually replaced by Indian capital. In fact, there was a lot of craziness among the Indians (including a good number of local investors) in making investment in the tea industry. That may be due to the fact that the persons with investable surplus had very limited options before them for investment. On the other hand, the tremendous success of the tea industry in Cachar in the nineteenth century inspired speculative endeavours. The official reports engendered the most extravagant belief in tea as a money spinner. It was thought that any one could run a tea garden. The investors were indeed reckless in purchasing land at exorbitant prices without considering the economic

viability of tea production. During the period from 1929 to 1949 there were about 110 tea estates in Cachar, out of which 56 gardens were owned by Indians. Out of the total acreage under tea of about 75000, the Indians-owned estates represented barely 22%. Out of 56 gardens under Indian ownership not more than half a dozen estates with a total registered acreage of about 1200 were Indian enterprises in origin. The rest were purchased by the Indians from the Europeans when the latter found it non-remunerative to hold any longer. There were instances where the purchasers paying only 10% of the purchase money got the balance from banks paying interests at rates varying from 6% to 9% per annum. The purchases of gardens at inflated (in fact, 'fancy') prices with bank money were greatly responsible for financial difficulties for most of the Indian owners in taking proper care of the gardens. Banks not only charged higher rate of interest but also did not advance adequate fund and hypothecation to run the garden properly, not to speak of making improvement thereon. As a result, years after years the quality of the gardens deteriorated. What was more, in their anxiety to repay the bank loans or realize return of capital invested, the Indian owners extracted the maximum in minimum of time even at the cost of the health and life of the plants or up-keep of the soil. While the Indian owners generally suffered from lack of adequate finance, their European counterparts had little scope to complaint on want of fund. There was a noticeable reluctance on their part in further investment for reasons best known to them (Cachar Plantation Committee Report 1950 : 71). In some quarter this is attributed to transfer of affections, while others opined political compulsion and a feeling of uncertainty. All these together caused fall of production years after years and gradually it turned out to be a losing proposition to run the gardens. Taking advantage of the Tea Control Act (1933) and the ready market for 'anything' produced, the investors generally did not loose and they some how managed to get back the money invested with good margin of profit. But to this valley it turned out to be a great loss. The valley which was a leader in tea production in all respects during the beginning of tea production in Assam started lagging much behind the other districts of Assam. It got the stigma of the region of 'uneconomic tea gardens'. This state of affairs left this valley at the end of first half of the twentieth century with as many as 16 uneconomic tea gardens. They were: Arcuttipore, Binnakandy, Binodini, Silcoorie, Poloi, Kuttal, Singalta, Lallamukh, Chorgola, Singhacherra and Maguracherra, Kallaincherra, Tuillak, Binnakandy, Borakai, Bora Jalinga, Rosekandy, Goombira, and Iraligool [Cachar Plantation Committee Report 1950 : 3].

In reference to the above and the remarks of Edger [Griffith, 1967 : 99] one question becomes pertinent – whether the European planters were more judicious in taking decisions for their estates? It goes without saying that the owner of a tea estate should be both industrious and judicious while dealing with one's garden. A decision in haste may mean some minor loss of capital to him, but that may result into loss of livelihood for the garden labourers who are already marginalised and dedicated to the garden for generations. When a garden is 'closed' or 'abandoned', it does not just deprive the labourers of their daily income but also access to medical and other benefits like drinking water and electricity supply. The tea garden labourers generally live in an 'enclave society' even after a garden is closed. They prefer to live within the garden, and do not go out for alternative employment.

The present scenario of the tea industry of Barak Valley does not offer an impressive account of the indigenous entrepreneurs. At present there are 29 (out of 102) gardens under such ownership in the valley. Out of these 29 gardens, except Rosekandy and few others, all others are uneconomic or closed or in the verge of closer. Though the major constraints for the development of tea industry in particular and the economy in general in the Barak Valley are factors like communication, power supply, etc, this presentation would like to reflect on the indigenous entrepreneurial ability and the mind-set of the people which are also relevant factors for the economic development of a region. The entrepreneurship, as a 'stand alone' factor, will certainly not be able to solve all the problems but could help in minimising the problem within the existing structure. The example of Rosekandy Tea Estate may be cited here. This garden was uneconomic in 1950 and also in 1985, but after that, by making serious effort, revamping the management, it has reached to a level where the average productivity is almost at par with the top tea estates of India.

3) **Capital Investment:** Inadequate capital has been found to be one of the major causes of poor performance of tea industry in Barak Valley. The existence of higher percentage of unproductive (above 50yrs.) tea bushes in tche valley (table-3) exhibits less expenditure on plantation and replacement of old bushes. Higher the age of Tea bushes, poorer would be the quality of tea and lesser would be price fetched in the market. But even today, despite recession in the market for tea, good quality tea still fetches lucrative prices in both national and international markets.

Table 3
Age Group of Bushes during 1978-1993 in Barak Valley

Age of tea bushes in years	*Barak Valley*				*Brahma-putra Valley*	*Assam total*	*All India*
	1978	*1983*	*1988*	*1993*	*1993*	*1993*	*1993*
0-5	6.4	7.2	7.5	8.7	7.2	7.5	7.7
5-10	7.2	8.4	9.0	11.3	10.6	10.7	9.9
11-20	12.1	11.5	10.8	14.6	14.8	14.8	13.4
Subtotal (0-20)	25.7	27.1	27.3	34.6	32.6	33.0	31.0
21-30	7.6	9.2	11.9	11.1	14.7	14.2	13.4
31-40	8.7	7.0	8.4	9.3	13.1	12.5	11.9
41-50	10.3	10.8	9.5	6.8	8.9	8.6	8.0
Subtotal (21-50)	26.6	27.0	29.8	27.2	36.7	35.3	33.3
Over 50	47.6	45.9	42.8	38.1	30.6	31.7	35.7
Grand total	100	100	100	100	100	100	100

Source: *Statistical Profile of Barak Valley 1998.*

If a tea garden is maintained properly it gives good returns but the owner needs to have some patience. Tea gardens cannot give quick returns like share markets but when it starts giving profits that continues for long and with certainty. Thus an owner of the garden needs to be a genuine investor, and not a speculator.

4) **Infrastructure:** Although we are discussing various aspects of tea industry in Barak Valley as an isolated case, it is necessary to appreciate that no sector in an economy is independent. Neither any sector can develop in isolation nor can it take the whole economy forward alone if it does not get the support of basic

infrastructural facilities like communication (road, rail and air), power supply, good governance, favourable business atmosphere, etc. The pre-condition for the flow of investment in any region is the availability of basic infrastructure. Capital in a free and globalise economy is always very mobile. It is withdrawn from the place where margin of profit falls or becomes uncertain and moves to that place where the expectation of the same becomes higher. So, while endeavouring the repetition of the nineteenth century like economic situation today in Barak Valley we must make a relative assessment of our present infrastructural condition with respect to that golden time. This region at that time was well integrated with the mainstream economy and the mainland through water and road transport. The length as well as duration of journey between the mainland and this region during that period was much less than that of now. Even then the British rulers at that time did not consider those infrastructures sufficient for business activities and made the region connected by rail transportation. These prove that beside the availability of natural resources, efficiency level of the entrepreneurs and funds awaiting investment, the requisite infrastructural facilities are also basic for the economic development of a region. In the present stage of economy, there are plenty of opportunities for the investors to invest; so funds are withdrawn from the complex zones and invested in the convenient ones. This trend is already operational in the economy of Barak Valley, and that is not unusual in economic terms. The present status of infrastructural facilities in Barak Valley - either in respect of road transportation or rail transportation or electricity and their regularity - hardly needs any further elaboration.

Tea industry in India as a whole is passing through a crisis period and the government is providing various relief measures to bail them out. However, unless the infrastructural facilities— both intra and inter regional – are developed it would be difficult for tea industry in Barak Valley to survive. On the other hand, if tea industry gets the requisite infrastructural and entrepreneurial support it would certainly prove itself again as a leading sector in taking the whole economy of Barak Valley much forward. An analysis reveals that if only the 28 presently sick gardens are revived, there will be direct release of additional 3320550 kgs tea, additional income of more than rupees 22 crores, and 5000 new employments (annexure-II and annexure-III). There would also be other indirect benefits through various forward and backward linkages.

Conclusion

The presentation would like to conclude by reiterating that there are huge potentials in Barak Valley to take her economy to a new height, provided some reasonable amount of infrastructural facilities are made available for the development. The second priority should be the development of indigenous entrepreneurships and good governance. A critical analysis of facts raises some questions: Have we done so far our best in exploring the available resources? Are we making best utilisation of the centrally sponsored schemes so far released to us? Why the formal trade undertaken so far with the neighbouring countries are not in our favour? Should we opt only for quick profits? What is our saving-investment ratio? Are we making over-expenditures, or keeping something for capital formation? The answers to these and similar more questions may demand a change of the mindset on the part of all concerned in the development activities, including the common people. Though the infrastructural inadequacy has been correctly held as the major cause of backwardness, we would give equal weightage on the above mentioned facts in order to create a ground for inclusive and sustainable development of Barak Valley. Otherwise, if at a point of time the infrastructure improves and the region develops, that development will not be inclusive one. The region will develop but the people will remain poor.

Annexure-I

Views of Thomas Fisher, First Superintendent of Cachar, on suitability of Cachar for tea cultivation

"In regard to climate it would appear that neither the heat nor cold are as intense in Cachar as in China but the rainy season would seem to occur at precisely, in the same time in both countries, the rivers in Cachar being swollen from February as in the tea districts. The conditions of the atmosphere accompanying the blowing of certain winds appear to be precisely similar in both tracts. The latitude of Man-choo-foo one of the tea districts, is the same with that of much of Cachar. The face of the country appears similar, there being both mountains and hills, the bases of which are worked by numerous rivers and lakes flowing through river tracts with gentle falls to the sea. The soil on the slopes near the foot of many of the hills and mountains is such as Dr Abel states of the tea plant to be grown upon; viz., loose and gravely formed by the debris of sandstone, silts and granite and containing very little vegetable mould."

[Source: Letter from T Fisher, Esq., AAGG, to G S Gordon, Secretary, Committee for Tea Culture, London, dated 25th March, 1834].

Annexure-II

Area Extension and Employment Generation from Revival of Sick Gardens

Sl. No.	*Code name of the sick gardens*	*Area for extension (hec)*	*Additional employment available*
1.	SG1	300.00	672
2.	SG2	7.66	17
3.	SG4	4.05	9
4.	SG5	200.00	448
5.	SG6	260.65	584
6.	SG9	100.00	224
7.	SG16	49.80	112
8.	SG17	85.00	179
9.	SG19	103.49	232
10.	SG23	43.20	97
11.	SG25	342.85	768
12.	SG26	30.00	67
13.	SG27	121.46	272
14.	SG28	350.00	784
Total		1998.16	4465

Source: Tea Board of India, Individual survey by the researcher

Annexure-III

Expected Benefits from the Revival of Sick Gardens in Terms of Production and Income

		Present					
Sl.	*Code name of the garden*	*area under produ-ction (hec)*	*Produc-tion (kg)*	*Average prod-uctivity (kg/hec)*	*Expected produc-tion after revival (kg)*	*Net gain after revival (kg)*	*Addl. Income generation at price Rs.68/-*
1.	SG 1	287	334776	1166	652064	317288	21575584
2.	SG2	430	318143	739.86	976960	658817	44799556
3.	SG3	98	19342	197.36	222656	203314	13825352
4.	SG4	377	326850	866.97	856544	529694	36019192
5.	SG5	327	189332	579	742944	553612	37645616
6.	SG6	125	21434	171.47	284000	262566	17854488
7.	SG7	155	80482	519.23	352160	271678	18474104
8.	SG8	32	8000	250	72704	64704	4399872
9.	SG9	250	189750	759	568000	378250	27721000
10.	SG10	227.88	151364	664	518016	366652	24932336
11.	SG11	52	37224	716	118144	80920	5502560
12.	SG12	153	73788	482	347616	273828	18620304
13.	SG13	238	186773	785	540736	353963	24069484
14.	SG14	135	351	2.60	306720	306369	20833092
15.	SG15	308	31416	102	699776	668360	45448480
16.	SG16	571	487225	853	1297312	810087	55085916
17.	SG17	69858	684747	9.80	1585856	901109	61275412
18.	SG18	500	178463	356.92	1136000	957537	65112516
19.	SG19	325	152009	467.72	738400	586391	39874588
20.	SG20	818	628553	768	1858496	1229943	83636124
21.	SG21	43	16106	374.56	97696	81590	5548120
22.	SG22	49	663	13.53	111328	110665	7525220
23.	SG23	NA	NA				
24.	SG24	73	29193	400	165856	13663	929084
25.	SG25	145	44829	309	329440	284611	19353548
26.	SG26	24	4888	203.66	54528	49640	3375520
27.	SG27	159	119216	750	361248	242032	16458176
28.	SG28	44	16673	378.90	99968	83295	5664060
	Total	6643	4331590	652	1011040	3320550	225797400

Source :Tea Board of India, Silchar

REFERENCES

Awasti, R.C (1968), *Economics of Tea Industry,* United Publishers, Guwahati.

Bhattacharjee, J B (1977), *Cachar Under British Rule in North East India*, Radiant Publishers, New Delhi.

Cachar Plantation Committee Report (1950), Ministry of Labour, Government of India, New Delhi.

Digest of Tea Statistics (of various years), Tea Board of India, Calcutta.

Economic Survey of India (relevant years), Ministry of Finance, Government of India, New Delhi.

Gait, E A (1981), *A History of Assam*, Lawyers' Books Stall, Guwahati (First Published, 1905).

Griffith, P (1967), *History of Indian Tea Industry*, Weidenfold and Nicolson, London.

NCAER (1961), *Economically Weak Tea Gardens in North East India*, New Delhi.

NCAER (1962), *Techno-Economic Survey of Assam*, New Delhi.

Report of the Plantation Enquiry Commission (1956), Govt. of India, New Delhi.

Sample Survey of Unemployment Situation in Tea Gardens of Assam (1958).

7

Role of Banking in the Development of Agriculture in Barak Valley

Sudipa Rakshit

The World Bank in India (November, 2006) observes "Despite some progress in a few pockets, overall economic growth in the region has remained low, poverty is high – especially in Assam, as is the degradation of natural resources. The region's agriculture sector has been declining, and diversification into services and manufacturing has been inadequate. All in all, then, the region continues to miss out on the economic growth acceleration witnessed in much of India." According to the *Economic Survey of India 2007-08*, "Growth is of interest not for its own sake but for the improvement in public welfare that it brings about. Economic growth, and in particular the growth in per capita income, is a broad quantitative indicator of the progress made in improving public welfare". It is to be mentioned that during the decades of 1950's and 60's Barak Valley was one of the richest regions of Assam, but the present scenario of Barak Valley is one of the low income areas of Assam. Cachar is the largest district of Barak Valley. The total population of Cachar was 14,42,141 as per 2001 census, of which rural population constituted 86.06% and the population Below Poverty Line (BPL) was estimated at about 46% [PLP 2008-2009] of total population which was far above the all India average at 28.0% during 2004-05.

The main reasons of incidence of poverty are due to very week infrastructural facilities, both social as well as economic infrastructure. There is shortage of basic infrastructure for agriculture – like irrigation, road, electricity, storage facilities and marketing - in the district. In view of the resource endowment of the region agriculture led growth strategy appear to be the best bet. With the increasing liberalization and globalization of Indian Economy in general and agriculture sector in particular, the agricultural sector needs to be more responsive and competitive to face the external challenges. This would necessitate huge investments in infrastructure, technology, post-harvest management, research and development, extension services, etc. [PLP 2008-09]

Agriculture is largely dependant upon outside finance for the development. In fact, the lesson of universal agrarian history is that an essential need of agriculture is credit. The agriculturists must borrow due to the fact that an agriculturist's capital is locked up in his lands and stock. Hence, credit is not only essential but inevitable and as such it is not necessarily objectionable, nor a sign of weakness [Nicholson, cited in Ghosal, 1972]. However, agriculture relies on available resources of various kinds which can be referred to as 'capital', and these resources include :

Financial capital can be considered as the nucleus of all these capitals. Credit is an important input in the production process. Myrdal has termed credit as an accelerator of development [Joshi, 2002]. Insufficiency of institutional credit in rural areas of Barak Valley is seriously impeding growth of agriculture in the region. Agriculture is the most important area in which the Cachar district has an inherent strength to achieve full self-sufficiency. But the full utilization of the potentialities and achievement of self- sufficiency are not possible without properly addressing the issues, such as, viability and sustainability of agricultural lending - particularly to marginal and small farmers, operational efficiency, recovery performance of the banks, small farmer coverage under bankable scheme, balanced sectoral development, etc.

Table 1

Strategies for building up various forms of capital required for agricultural food production

Natural Capital	*Financial Capital*
• Water harvesting, water management	• Stable Markets
• Soil conservation	• Subsidiary activities
• Biological pest control	• Readily available credit
• Composting, manuring	• Post-harvest technological opportunities
• Diverse systems – many types	• Value-added activities
• Conserving genetic resources	
Social Capital	*Physical Capital*
• Cooperatives	• Improved tools, machinery
• Extension workers: Government, methods NGO, private	• Precision agriculture
	• Low dose sprays
• Farmers self-help and research activities	• Improved crop varieties
	• Roads
• Social values and systems	• Processing Plant
Human Capital	
• Improved nutrition	
• Education	
• Health	

Source : Based on information contained in *Pretty (1999)* *Cited in* Vanloon *et al 2005, p. 48.*

The climate and soil of Barak Valley in general and Cachar district in particular, have all along been congenial for cultivation of a variety of crops, rice being the most important one. Barak Valley possesses outstanding potentials in agriculture in the following fields :

- It is climatically favourable for cultivation of important agro-economic species as grown in other parts of the state.
- The district has the large acreage of water bodies of huge potential which need to be tapped.
- The district enjoys higher productivity of paddy per hectare than the state in almost all the years since 1990-91.
- The prospect of increased productivity with the adoption of improved technology is also very high.

The agricultural sector of the Barak Valley is covered under one or other category of Fundamental Backwardness. The following fundamental backwardness exists in this vital sector of the district, viz.,

i] Subsistence farming in most of the AEO circles.

ii] Lack of Irrigation facilities.

iii] Backward technique of production.

iv] Poor rural infrastructure.

v] Chronically flood affected cultivated areas.

vi] Abysmally low bank credit inflow into the agricultural sector.

As a result, the possible externalities and spread effects could not be generated in this sector and the growth impulses in the sector remains low. [Rakshit, 2006]

Agriculture occupies a vital position in the district's economy, engaging about 37 per cent of the total working force. The district enjoys a higher yield level in the production of paddy per Ha than the state in almost all the years since 1990-91 (Table 2), but total production is not sufficient.

Table 2
Rice Yield during the period 1990-91 to 2003-04 in Cachar, Assam and India

Year	*Rice Yield (Kg/Ha)*		
	Cachar	*Assam*	*India*
1990-91	1752	1313	1740
1991-92	1690	1290	1751
1992-93	1770	1308	1744
1993-94	1180	1331	1888
1994-95	1823	1350	1911
1995-96	1778	1421	1797
1996-97	1959	1334	1882
1997-98	1595	1356	1900
1998-99	1854	1344	1921
1999-00	1758	1483	1986
2000-01	2132	1568	1961
2001-02	2201	1539	1913
2002-03	2093	1488	1862
2003-04	2149	1534	–

Source: *1. Deputy Director, Economics and Statistics, Silchar, Cachar.*
2. Economic Surveys. Assam, GOA, various issues
3. Economics Surveys, India, GOI, various issues.

The Barak Valley, which was a rice surplus zone till 1960-61, has been converted into a rice deficit zone since 1971. And for the district of Cachar the deficit amounts to 26,000 MT annually [Agriculture Status Paper, Agriculture Department, Silchar, 2000, Cachar]. The deficit is being met by importing rice from other regions of the country. Being endowed with rich fertile land and water resources, it is paradoxical that the district is incurring huge outgo of revenue by importing rice and vegetables from outside. In addition, the fact that the locally produced rice/vegetables gains better price than those imported from outside the district offers greater scope for development.

Over the 14 years period from 1990-91 to 2003-04 under the present study, the percentage of direct agricultural lending was averaged at only 6.3 per cent in Cachar, which has been well below the target of 18 per cent of Net Bank Credit for agriculture. Thus, there exists vast potential for agricultural lending and it is observed that the actual agricultural lending since 1991 appears to be inconsistent with the potential demand for agricultural credit in the district. If adequate credit and assured irrigation can be extended, the present cropping intensity of the district which is 123.5% [PLP 2008-09] can be easily raised up to 150%. The very sizeable need for increasing credit supply for agriculture in the district can be gauged from the fact that the target of Rs. 2334.25 lakhs (total agricultural credit) for 2006-07 could not be achieved. Only 56.5 per cent of target had been achieved [AAP, Lead Bank, Silchar, 2007-08].

In the present study, an attempt has been made in making projections of the potential credit requirements in total paddy crop area in the district. A vast credit gap to the tune of 99 per cent exists in the district in terms of 'potential demand'. The situation in other years is almost similar. As Cachar is largely a peasant economy having 93 per cent of small and marginal farmers (including landless farmers also), this problem assumes a critical significance. So the various constraints faced by the farmers as well as by Commercial banks need to be problematised and addressed rigorously for an in-depth analysis. The present study therefore is an attempt to make an in-depth analysis.

The per capita agricultural credit is found to be equally low at Rs. 27.09 in the district of Cachar during 2003-04.

Table 3 (A)
Working Capital requirements in agriculture in Cachar District (2003-04)

Name of major crop	*Area under different crop (Ha)*	*(Bigha)*	*Scale of finance (Per bigha Rs.)*	*Total working Capital requirements (Rs. lakh)*
Autumn Paddy (May-June/Aus)	9250	69375	2189.00	1518.62
Winter Paddy (Nov-Dec/Sali)	90081	675607.5	1817.00	12275.79
Summer Paddy (April-May/Boro)	9865	73987.5	2044.00	1512.30
Total	109196	818970		15306.71

Source: 1. Status Paper of Cachar district (2002-04) Office of the DAO, Cachar, Silchar (1 Ha=7.5 bigha).

2. Scale of finance (per bigha) for different crops approved by D.L.T.C. Cachar, Silchar.

Table 3 (B)
Estimated credit gap for crop loan of KCC holders in Cachar District (2003-04)

Total working capital requirement for crop loan (as per scale of finance) (Rs. in lakh)	*Total crop loan supplied through KCC*	*Estimated credit gap*
Rs. 15306.71	156.10 (1.01%)	15150.61 (99%)

Source: Annual Credit Plan (2003-04) Lead Bank, UBI.

It may be mentioned that the share of credit flow to agriculture sector (including crop loan) was 15.4 per cent, 14.5 per cent and 12.9 per cent respectively indicating concentration of financing on "Other Priority Sector" activities. Though the crop loans through KCC have ostensibly been introduced in the district during 1999-2000, the impact of the scheme is not yet felt in actual practice. Per hectare crop loan is very much low as shown in Table 4. The principle of seasonality in lending and recovery of crop loan, though good in principle, has yet to be popularized in practice.

Table 4
Area under rice, total production of rice and crop loan (1998-98 to 2003-04)

Year	*Area under Rice (in Ha)*	*Total Production of Rice (in tonne)*	*Crop Loan (In Rs. lakh)*	*Crop Loan (In Rs. per Ha)*
1997-98	103478	165074	3.12	3.01
1998-99	102941	190864	5.61	5.45
1999-00	103464	181850	1.78	1.72
2000-01	104000	221685	10.71	10.30
2001-02	102093	224744	11.20	10.97
2002-03	105120	220058	86.30	82.10
2003-04	109196	234639	156.10	142.95

Source: 1. DAO, Silchar, Cachar.
2. Office of Deputy Director, Economics & Statistics, Silchar.
3. UBI, Lead Bank, Silchar.

Performance of the Commercial banks in respect of the (Reserve Bank of India) norms

The RBI prescribed certain norms which have to be followed by the Commercial Banks. The targets set regarding agricultural lending are as follows:

1. Credit Deposit Ratio – 60 per cent
2. Priority Sector Advances – 40 per cent
3. Advances to agriculture – 18 per cent of Net Bank Credit

Credit Deposit Ratio (C.D. ratio)

"The credit-deposit ratio is an important indicator of the degree of involvement of banks in lending" (10th Five Year Plan, Vol. II, p. 521). The Credit deposit ratio provides an insight into the spread of banking facilities, credit opportunities, trend and pace of development taking place in the states of the country.

The C.D. Ratio in the district at 26.45 per cent as on 31/03/2004 is very much on the lower side as compared to all India position of 58.7 per cent in view of the growth rate in deposits and also keeping in view the potential available for credit deployment. C.D. ratio of Commercial Banks as per sanction in India, Assam and Cachar is shown in Table 3.8 during 1990-91 to 2003-04.

Table 5
Credit deposit ratio in India, Assam and Cachar (in percentage)

Year	*India*	*Assam*	*Cachar*
1990-91	69.18	49.50	39.41
1991-92	61.01	49.00	35.45
1992-93	60.51	43.80	34.92
1993-94	56.62	39.80	31.24
1994-95	59.20	38.96	30.20
1995-96	59.80	40.30	28.08
1996-97	56.80	36.30	24.64
1997-98	55.30	33.60	22.57
1998-99	54.80	32.00	23.37
1999-00	56.00	32.00	20.03
00-2001	56.70	32.10	21.69
2001-02	58.40	31.70	22.00
2002-03	59.20	29.30	23.00
2003-04	58.70	30.81	26.45

Source: (1)1991-2004, Banking Statistics, quarterly handout, RBI (various issues).

(2) ACP, Lead Bank, UBI, Silchar, Cachar (1990-1 to 2003-04)

The C.D. ratio of Assam in general and Cachar in particular have been worked out at 30.8 per cent and 26.45 per cent respectively as on March 2004, which are very much on the lower side compared to all India ratio of 58.7 per cent. In view of the growth rate in deposits and also keeping in view the potential available for credit deployment, it is necessary for banks to identify new areas where they can invest fund. The overall CD ratio of the banks in the district has been improving during the last three years. However, CD ratio at 42.76% as on 31.03.2007 was very low. The performance of Co-operative Bank was found to be better at 62.42% as on 31.03.2007.However, it was mainly because of accumulated overdue loans made in the past.

Direct Finance to Agriculture

In the case of *direct finance* to agriculture also, the Commercial Banks could not attain the *RBI stipulations of 18 per cent lending of Net Bank Credit to agriculture.* Although the outstanding amount of agricultural advances has been going up in case of commercial banks, their share in bank credit has not changed appreciably over the last 14 years (Chart 1).

The chart 1 reveals that during the entire period of survey i.e. from 1990-91 to 2003-04 the ratio of direct finance to agriculture is varying from 5.93 per cent (1990-91) to 6.7 per cent (2003-04) which is far below from satisfactory. And this failure to achieve the stipulated percentage is reflected at all levels i.e. in national level (15 per cent), state level (in Assam it was 12.7 per cent during 2000-03) as well as in district level (Cachar). *None of the agencies in the district could achieve target prescribed for advances to agriculture sector and it was 9.41 % only during 2006-07 against the prescribed norms of 18%.*

Chart I - Outstanding Agricultural Advances of Commercial Banks Share in net bank credit (in percentage)

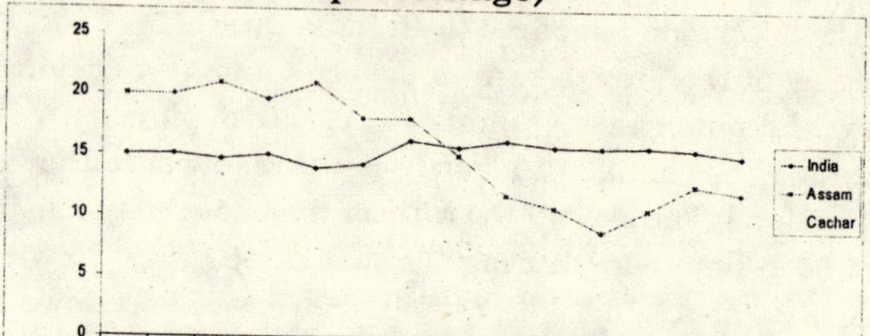

	1990-91	1991-92	1992-93	1993-94	1994-95	1995-96	1996-97	1997-98	1998-99	1999-00	2000-01	2001-02	2002-03	2003-04
India	15	15.1	14.7	15	13.9	14.3	16.3	15.7	16.3	15.8	15.7	15.8	15.6	15
Assam	20	20	20.9	19.6	20.9	18.1	18.1	15	11.8	10.7	8.8	10.6	12.7	12
Cachar	5.4	6.7	8	8	7.3	7.5	8.4	7.8	6.3	5.6	4	5	6	6.7

Source : *(i) ACP, Lead Bank, Silchar (various issues).*
(ii) Economic Survey GOI (various issues)
iii) Economic Survey GOA (various issues)
iii) Reports on Currency & Finance, RBI, 1990-91 to 1997-98

Note: i) The target for advances to agriculture is 18% of NBC for the domestic banks.
ii) Indirect finance to agriculture is taken only to the extent of 4.5% of NBC for computation of agricultural lending.

Thus closer examination of the functioning of commercial banking structure in rural areas of the district reveals that public sector banks have been generally conservative and traditional in their approach and have failed to show awareness of their social responsibility and to meet the expectations raised by nationalization.

Despite repeated efforts, the flow of credit to the agricultural sector failed to exhibit any appreciable improvement mainly due to the fact that Commercial Banks were not tuned to the needs of and requirements of the small and marginal farmers.

II
Micro level Study

R.S. Sheshadri [2008] observes "Farming at today's prices is becoming unviable for the small and marginal grower". This is particularly true for the small, marginal and landless farmers of Cachar who constitute 93% of farm families. It calls for quick step to ensure farmers remain interested in agriculture. One immediate measure to redress the situation is to gear up the banking structure to be instrumental in strengthening the agricultural infrastructure. After 17 years of economic reforms why no trickle down of agricultural growth has occurred is a matter of concern.

The present study was undertaken to examine the growth and constraints of crop loan provided through KCC by nationalized commercial bank including Cachar Gramin Bank (RRB) in Cachar district during 1999-00 to 2004-05. As Cachar shares the same agro climatic conditions with other two districts of Barak Valley, namely, Karimganj and Hailakandi, and infrastructural gaps are also similar, the findings of this study can be safely extended to relate the entire Barak Valley region.

Table 6

An overview of flow of crop loan since 1999-2000 (Cachar District) through KCC Scheme

Year	*Kisan Credit Card (KCC)*			*Kisan Credit Card (Financial) (Rs in lakh)*			
	Target (Bank)	*Sponsored (Agr.)*	*Issued (Bank)*	*Target (Bank)*	*Sponsored (Agr.)*	*Sanctioned (Bank)*	*Disbursed (Rs.)*
1	*2*	*3*	*4*	*5*	*6*	*7*	*8*
1999-2000	**950**	**294**	**60**	**46.65**	**20.34**	**1.78**	**1.78**
2000-2001	**998**	**1252**	**215**	**52.00**	**22.38**	**10.71**	**10.71**
2001-2002	**911**	**725**	**272**	**59.60**	**37.25**	**13.60**	**11.20**
2002-2003	**4460**	**5252**	**1180**	**139.34**	**756.92**	**192.48**	**86.30**

(Contd...)

1	*2*	*3*	*4*	*5*	*6*	*7*	*8*
2003-2004	6558	5624	1830	372.40	997.86	217.34	156.10
2004-2005	10700	6200	3859	1070.00	225.0	210.0	625.17
2005-06							690.63
2006-07							557.14
2007-08	10,525		4746				

Source: *(i) Potential Linked Credit Plan (2002-03 to 2006-07, 2004-05, 2005-06,2008-09).*
(ii) Lead Bank Office, UBI, Silchar.
(iii) Agriculture Department (Silchar)

It is seen from table that number of KCC holders are significantly less than non-KCC farmers in the study circles.

For the purpose of intensive study about the trend and pattern of agricultural finance in the district of Cachar, 280 farmers have been selected from 20 villages covering the entire district, by judgment sampling method.

An interim assessment of the working of the KCCs scheme from 1999-00 to 2004-05 in Cachar district reveals that there are **two edged problem** regarding the Credit delivery System—(a) poor coverage of farmers as well as cropped area. (b) the disbursed loan amount is inadequate to satisfy the farmers' requirement which is shown in Table 7.

The commercial banks have the national obligation of a steady and substantial hike in credit outflow. But experience has proved that it would be extremely difficult for the commercial banks at the macro level to attain the desired growth rate of around 18% direct credit flow to the agricultural sector due to unsatisfactory performance of the agricultural sector barring a few years during the 90's. For small and marginal farmers, the deceleration in the credit disbursal has been the maximum in the 1990s. Small and marginal farmers, thus, continue to be both credit and demand constraint [RBI 2000-01].

Table 7

Distribution of KCC farmers in surveyed villages in Cachar and perceived inadequacies of loan amount

Agrl. Subdivision	*Silchar*	*Sonai*	*Lakhipur*	*Total KCC holders*	*Total no. of surveyed farm household*
Perceived inadequacies of loan amount :					
Sufficient	0	0	0	0	
Moderate	4	3	1	8 (16.7%)	
Inadequate	6	20	14	40 (83.3%)	
No. of KCC Holders	10	23	15	48(100%)	280

Source: Field Survey during 2003 to 2004.

Note: Figures in the parentheses represent percentage to total KCC holder.

Whereas the banking credit outflow could gain a strong footing in southern region and in many other states; it is not very successful till now in North Eastern Region. It is rather dismal in this region [10th Five Year Plan]. In Cachar district it is more acute. So the question remains to be addressed as to what are the inherent defects in the structure or in the system of delivery of credit flow in the district.

The district of Cachar has three agricultural subdivisions viz., Silchar, Sonai and Lakhipur. The District may be called a peasant economy having 54.5% small and marginal farmers and a smaller size of operational holding of 0.25 Ha [Agricultural Status Paper 2002-03]. Landless farmers constitute another 38.5% of total farm families. **A peculiar thing is that most indigent farmers are rarely provided with the KCC while those who get the KCC do complain that the amount provided is not at all adequate to their actual need.** So the various constraints faced by the farmers as well as by Commercial banks need to be problematised and addressed rigorously for an in-depth analysis. Out of total 280 farmers interviews only 48 (around 17%) were found to be KCC holders.

Field study reveals some basic features of the agricultural sector of Cachar district. About 60% farmers are producing mainly for home consumption. Only a small marketable surplus is left. Over a large part of rural Cachar there is only one season farming. 75% of land is under mono crop. There is a fairly regular transfer of labour on a seasonal basis from the farm to the non-farm sector (Table 8).

Table 8
Off farm activities of the sample farmers of Cachar District (2003-04)

Agr. Sub-Div.	*No. of F.H.H.*	*No. of full time farmers*	*Service*	*Fishing*	*Other small business*	*Daily Labour*
Silchar	117	44 (37.6%)	10	39	40	17
Sonai	92	34 (37%)	16	07	17	18
Lakhipur	71	42 (59.2%)	04	0	25	0
Total	280	120 (42.9%)	30	46	82	35

Source: Field Survey

In the opinion of 60% farmers today agriculture is not at all a profitable activity, it is just a way of living. It is also found that in spite of prevalence of monetary economy in present day, barter economy still operates in the interior villages of the study area, which speaks about the backwardness of these villages. The consumption goods like onion, dry fish, vegetables, pulses etc. are exchanged against paddy. Generally, the rate of exchange is 2 kg paddy = 250 gm. onion or 100 gm. dry fish. Here in these villages 'for these farmers paddy plays the role which money plays for non-agricultural producers and consumers.'

It is worth mentioning here that being compelled by acute poverty, some marginal and small farmers resort to 'distress selling' of the upper layers of their farm land to the owners of Brick Kiln at a nominal rate resulting not only in less productivity but also in permanent land degradation rendering the farmers perpetually poor.

One of the important findings of the field survey is that over a large area of rural Cachar subsistence farming is in practice. A large proportion of the paddy produced is consumed directly. 13.2 percent of sample farms household found to be displaced persons from erstwhile East Pakistan who were settled in waste cultivable land and though they were in possession of the land since 1964 they are not given the status of land holders and revenue is collected as per '*Touzibahir*'[1] account. Though each agricultural family was allotted six bighas of land for homestead and cultivation, enjoyed the voting right, yet in terms of permanent holdings right they are not entitled to get the loan facility under KCC scheme. Refugees from erstwhile East Pakistan have introduced the cultivation of 'boro' paddy and 'ground nut' which they carry out extensively, failed to make it commercially viable due to lack of bank credit for various problems not beyond solution. Urgency of ready money in the hands of the farmers compels a large number of farmers in *distress selling* of paddy which is not recognized at the government level.

A vast credit gap to the tune of 99% exists in the district in terms of potential demand for the production loan. From the field survey it was found that out of 280 farmers (the study area) only 17.14% got KCC as shown in the following:

Chart 2

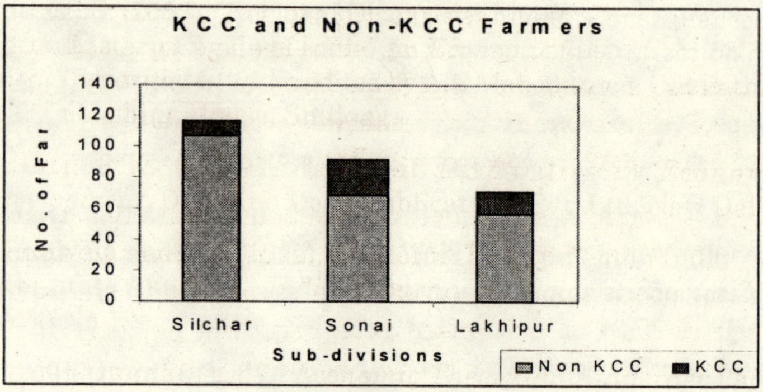

- Low level of ground level credit flow to the agricultural sectors can be attributed to two sets of factors—**(A) Demand side constraints** and **(B) Supply side constraints.**

Factors operating on the Demand Side for poor offtake of Bank Credit (KCC)

Field survey revealed that out of 280 sample farm households, only 48 farmers i.e. 17% of total sample farmers got crop loan under KCC scheme from commercial banks including Cachar Gramin Bank during 2003-04. The slow progress of KCC scheme and also poor off take can be attributed to both the demand constraints as well as supply constraints.

(A) Demand Constraints

The inadequacy of the agricultural bank finance may be due to a fusion of three sets of factors acting together as demand constraints which are shown in Table 9 below:

Table 9
Demand Constraints

Economic	*Behavioural*	*Societal*
a) Lack of Entitlement set	(a) Lethargy	(a) Lack of awareness
b) High transaction cost	(b) Self contentment	(b) Lack of education
c) Absence of title to the land	(c) Fear of punishment in case of non repayment	(c) Misinformation and or asymmetrical communication
d) High default rate		
e) Infrastructure gap etc.		
f) Traditional farming		
g) Inadequate Extension service network etc.		
h) A wide Informal Credit Market		

Table 10
KCC farmers and land holding (in bigha)

	0 to < 3 bighas (landless)	*3 to < 7.5 bighas (marginal)*	*7.5 bigha to < 15 (small)*	*15 (bigha & above)*	*Total KCC holders*
Silchar	0	3	4	3	10
Sonai	1	4	13	5	23
Lakhipur	0	1	10	4	15
	1 (2%)	8 (17%)	27 (56%)	12 (25%)	48

1 Hectare = 7.5 bigha
Source: As per questionnaire
Note: Classification of land holding as per Agriculture Department, Assam.

- **Lack of entitlement sets :** As is clear from table 10, access to the loan market is systematically interlinked with the size of the borrower's entitlement set. Out of 280 sample farm households 36.4% of the total farmers are unaware of the KCC scheme, remaining 63.6% though are aware of the scheme, only 27% of aware farmers got KCC during 2003-04. Out of them 25%

belong to big land holder, 56.25% are small land holders and 17% are marginal farmers and only 2% are landless farmers.

- **High Transaction Cost :** The transaction costs of lending and borrowing are a major barrier to providing access to microfinance services for the poor on a sustainable basis. For small loans these transaction costs can be burdensome in relation to the value of the loan itself. [McGuire & Conroy, 1997].

  Low interest on rural credit is a farce. The cost of small borrowers sometimes happens to be the prime lending rate plus 25%. Here in the district of Cachar apprehension of high transaction cost deters 18.5 percent of sample farmers approaching the commercial banks for agricultural credit. The cost of transaction on average is as follows (made on the basis of interview with the KCC holder).

Table 11

Transaction cost of KCC per account (as reported by the KCC holders) to be borne by the farmers

	Items	*Cost (in Rs.)*
1.	Photo	30.00
2.	Cost of getting Land Document Certificate from Circle Office	250.00
3.	G.P. Certificate (by Secretary / President)	10.00
4.	Cost for filling up KCC form	10.00
*5.	For making estimation of the project (cost)	200.00
6.	Cost of Judicial Stamp	40.00 (Rs 10 x 4)
7.	Cost of Revenue Stamp	4.00 (Re 1 x 4)
8.	Embossment Cost for Judicial Stamp	20.00
9.	Cost of opening bank account	100.00
10.	Transportation Cost (for going to circle office, G.P. Office, Bank) from date of application to disbursement of credit	200.00
	Total Cost :	864.00

* In case of Cachar Gramin Bank this cost is to be excluded as per statement of the beneficiaries)

Source: Field Survey.

Here the main components of borrower's transaction costs included documentation expenses, travel costs and incidents. And the opportunity cost of time spent in negotiating the loan has been overlooked. Borrower transaction costs were estimated at Rs. 864/- per loan account.

In the surveyed circles the total loan disbursement to the 48 KCC holders amounted to Rs. 4.32 lacs during 2003-04 (total credit). So taking Rs. 9000/- as the average loan per loanee farmer the transaction costs accounted for 9.6%, of the total credit which is too high. As many as 30 farmers (62.5%) out of 48 KCC holders reported that in practice the transaction cost was even much higher than the reported amount. The *consideration money* of Rs. 500/- per account is included in the cost estimated.

- **Traditional farming :** 40% farmers are doing local variety, 44% are not using chemical fertilizer or even insufficient quantity of organic manures, 53% farmers are not doing farming with power tillers and not using pump-sets for financial constraint. Due to lack of finance 28% farmers are not doing HYV seeds.

Table 12

Reasons cited by farmers for non-adoption of HYV seed – fertilizer technology

Reason cited	*No. of farmers*	*Percentage of total sample farmers*	*Rank*
1. Lack of finance	78	27.9	1
2. Lack of knowledge	38	13.6	4
3. Flash flood	63	22.5	3
4. Absence of Irrigation facilities	66	23.6	2
5. Risk - aversion	30	10.7	5
6. Other reasons	28	10	6

Source: Field Survey.

Note : The figures are not additive.

- **Lack of irrigation facilities :** The farmers of Cachar district suffers from a serious insufficiency of irrigation facilities. Actual total area irrigated by different sources of irrigation is only 0.32% of total cultivation area under 19 AEO circle of Cachar during 2002-03.
- **Inadequate Extension Service Network :** This is another contributory factor for low coverage of bank credit. 36% of farmers are deprived of the bank credit facility due to lack of knowledge. Out of 280 sample farmers about 56% don't know the Village Level Extension Workers of the respective areas. Non filling up of vacancies of VLEW has further created a vacuum in the extension services.
- **Informal Credit Market :** 48% sample farmers are found to be indebted to various non-institutional sources of finance despite proliferation of branches of PSBs in Cachar. Poor farmers often lack accessibility to institutional sources of finance and have to obtain loans at varying rates of interest.
- **The informal moneylenders** as found in this study are categorized as under:

(A) Registered – Professional

(B) Unregistered – Non-professional

(i) **Clandestine** (ii) **Apparent** & (iii) **Occasional**

The sources of credit are not always mutually exclusive.

The rural marketing scenario is characterized largely by the existence of inter-linkages in factor and product market. In Cachar, during survey period 29.4 percent farmers are found to be affected adversely by factor product* market inter-linkage. Similarly, the inter-linkages in the factor markets (i.e. between credit and labour markets) limits the technology adoption by the poor farmers, by way of putting extra-economic demand on farmer's labour at the crucial time, mainly during sowing and cutting the paddy. Thus this type of linkage was found to deprive 9 percent of loanee farmers (i.e. 12 farmers) in Cachar (i.e. sample farmers) in getting their actual wage. During field visits in the sample villages, prevailing systems (local) of agreement with regard to the land in exchange of hard cash were revealed. These understanding as regards transfer of land temporarily in lieu of money are always seen in the villages of Cachar. Needy land holders, thus get requisite amount of money for day-to-day uses. They do not bother to approach the commercial banks in order to avoid much formalities and wastage of time etc. There are various methods of mortgage through *Ekrarnama*[2], *Swaranlipi*[3], *Izara*[4], etc.

- The most predominant source of private credit in rural Cachar is rich landlord / agriculturalists, followed by servicemen which is gaining importance in some of the villages.

- The mode of repayment are various :–

(i) Crop to crop; (ii) Money to money; (iii) Money to crop; (iv) Money to Labour. The rate of interest of private debt is found to be ranging from 10% to 20% on monthly basis which is exorbitant.

Thus recurring need of credit which is a '***necessary evil***' at a high rate of interest restricts the peasant farmers power to generate a surplus from their agricultural holding.

To find out the relationship between bank credit and private debt sub-division wise Z-test is done. The values of Z come out to be greater than table value of Z @5% level of significance. The values tell that the number of private borrowers is more in all the three sub-divisions compared to the KCC holders.

It is inferred that ***expansion of the commercial banking couldn't lessen the private debt of the farmers in the Cachar district.***

(B) Supply side constraints

- **Asymmetric information:** Because of limited information, the commercial banks restricts its activities almost exclusively to rationing loanable funds to medium and big borrowers who are deemed more creditworthy. The bankers face this problem of asymmetric information, often known as a '*lemon problem*'.
- **Imperfect competition:** Due to limited competition in the rural banking sector, most of the PSBs are unwilling to accommodate new and unknown potential borrowers due to perceived risks of non-repayment of loan. This imperfect competition is identified by Nobel Laureate (2001) Joseph Stiglitz as one of the seven reasons of market failures in LDCs financial markets.
- The credit requirements of the poor farmers are for small amounts, required at very frequent intervals, for both consumption as well as for production purposes and at odd times which cannot easily be dealt with by the banks within the viability norm.
- **Lack of applicants:** According to some bank managers, the banks face the problem of achieving the target for lack of big farmers, inadequate numbers of loan application etc.
- **Exclusion Problem :** Ongoing Kisan Credit Card (KCC) is suffering from "exclusion problem". As per the "Status Paper 2002-2003" [Agriculture Department, Silchar, Cachar]. 72.5 percent households belong to the category of marginal farmers and landless

farmers. But only 18.7 percent of marginal farmers got benefit of this scheme during 2003-04 which does not reflect their share among the beneficiaries. Though KCC is admissible to every farmer with minimum land holding (having land clearance certificate), but the bankers and agriculture department both are not sure of asset generation by the farmers with such a less quantum of land, and as such generally the cases below 5 bighas are not entertained as was found from field survey.

- **Rejection of loan applications without any genuine ground** : About 24% prospective borrowers alleged that the banks rejected their loan application without sufficient and genuine reasons. This finding is in consonance with the observation made by General Manager, RBI in the SLBC meeting 2003. [IIBM, Guwahati].

- **Coordination failure:** In the absence of adequate coordination among different complementary investments at the same time, the effort to bring sustainable agricultural development could not succeed.

- **No-trust syndrome:** Bank credit suffers from the malady of 'no-trust syndrome'. Farmers find bankers unfriendly, while banker, on the other hand, perceive every small borrower as a potential 'non-performing assets.'

- **Low Level of Credit-absorption capacity in the district:** 'Lack of infrastructure facilities which are a hurdle in the way of augmenting credit facilities is very much conspicuous in the district. Infrastructure support in the form of forward and backward linkages for agricultural activities is not well developed.

- **High incidence of over-dues in Agricultural Credit:** One of the major stumbling blocks in enlarging credit to agriculture and allied activities is that of mounting overdues. The present study finds that 55% are defaulters among 154 borrowers out of 280 sample farmers. About 50% of them are 'Willful defaulters'.

  From the secondary data it is revealed that total overdue position of loan and advance is 57.25 % in respect of KCC during 2006-07. ***It was observed that due to poor recovery of loans disbursed in the past, banks were insisting on additional securities over and above security norms prescribed by RBI.***

  - ❖ Adoption of restricted lending policies due to fear of NPA provisions.
  - ❖ Poor monitoring, supervision and control by the controlling offices.
  - ❖ Excessive dependence on Govt. sponsored programmes under Agriculture and Allied Sector.
  - ❖ Lack of sound appraisal techniques and post-disbursement supervision .

- Field level government functionaries of line departments are promoting SHGs as they are given target by their authorities. .However, it is observed that they generally approach people with the element of subsidy available in the system.

- KCC farmers automatically come under the crop insurance scheme "The Rashtriya Krishi Bima Yojana" or National Agricultural Insurance Scheme (NAIS) which provides comprehensive risk insurance against yield losses viz., natural fire and lightning, hailstorm, flood, drought etc. Yet unfortunately the farmers of Cachar district are totally deprived of insurance facility for crop damage. The district witnesses frequent and widespread occurrence of floods almost every year, resulting large scale damages to property and standing crops. But the farmers are not apprised of the process how to place claim in case of damage of their standing crop, though premium is deducted at the time of disbursement of loan amount. For lack of transparency in the policy matters, the whole thing is in a standstill position. This factor also acts as a disincentive to many farmers for taking KCC for production purposes.

The findings in the district level (Cachar) corroborates to the observation made in the 10th Five Year Plan (2002-2007), that 'The progress of the scheme . . . is dismal in the north east.' The institutional credit cannot provide an alternative to the money lender till now in the district, but an addition to his dealings. One of the major causes of this failure is banks' *doctrinaire aversion* to provide consumption credit to cultivators.

As a whole credit delivery system has failed to attain the objective of augmenting the total volume of bank credit to the agricultural sector, more particularly to the small and marginal farmers which has been the focus of bank credit policy since 1999 i.e. inception of KCC.

III

Conclusion and Suggestions

In order to ensure required flow of credit to farm sector, Banks may initiate following measures:

- It has been noticed that the prospective borrowers are encouraged by the sponsoring agency for taking mainly crop loan. But the other potentialities in the agriculture sector where vast scope exist for future development, like fishery, poultry, piggery, horticulture should be explored and encouraged for credit financing purposes.

Diversification of the agricultural sectors is the need of the hour to make stride in the way of transforming subsistence agriculture to modern agriculture.

- 23% of the sample farmers in the study area are having a pessimistic outlook regarding the way of life they lead and there is nothing more powerful than individual incentives to break this. The motivation of the farmer is important at all levels of production i.e. at the time of preparing land, sowing the seeds etc. This factor largely depends on the extension agencies of the state government. Local NGOs should come forward to take active part in this task.
- Awareness creation programmes should be undertaken on a large scale by the banks in the villages. Time of meeting conducted by banks should be fixed at the convenience of the farmers and not taken during harvesting hours. Many motivation programmes turn into failure due to lack of peoples (farmers) participation. For effective management of the programmes, a concerted effort on the part of the Bankers, Agriculture officials and Panchayats are needed.
- Some schemes relating to off-farm activities should be introduced by Commercial Banks for the poor farmers for upliftment of the conditions of the poor farmers and adhiars as also for agricultural labourers. All over the district of Cachar, the cultivators are producing different components of plough, *uri / ura* (basket made of cane) etc. to cover up their disguised unemployment situation. They can be brought under bankable scheme under the effective supervision of extension agencies.
- Agro-climatic condition of Cachar district is congenial for development of the plantation crops viz. peanut,

cashew, rubber etc. In the district of Cachar about 6% of surveyed farmers (especially those who came from erstwhile East Pakistan) are found doing cultivation of groundnut but not on a commercial basis. The agricultural department and commercial banks should take necessary steps in the form of motivating the farmers and providing finance to the farmers to go for production of ground nut on commercial basis which can meet the oil seed deficit to the extent of 90% in the district to a large extent.

- Establishment of post-harvest management facilities, like cold storages, are the needs of the day in the district. At least 12 numbers of cold storage are needed in Cachar [Agr. Deptt.]. This will help to increase the shelf life and thus help to avoid the wastage of perishable products and ultimately will help in increasing the returns for each and everyone in the value chain. The banks should provide loan at low interest to create the required infrastructure in the main producing areas.

- The poor and needy refugees from East Pakistan were given allotment of land some 35-40 years ago over a large area of vacant waste land and hilly tracks. It is clear, these peoples are not to be treated as '**floating population**' since the days of allotment these poor people are continuously residing there and as such their names are also entered in the electoral roll as per rules. So they are also entitled to avail of the facilities provided by the Commercial Banks working nearby without questioning the admissibility of the cases, because of their non-possession of pattas. They are allottees as per govt. orders.

- The farmers can undertake some agro-based eco-friendly sub-system of agriculture. The farmers can undertake production of earthen glass and pot for consumption of tea, water etc. It can be done on a commercial scale, which would provide subsidiary employment opportunities to the small, marginal and even agricultural wage earners. Banks can also formulate bankable scheme to cover this potential area within lending parameters.

- There is problem of identifying professional moneylenders, since many of them invariably combined money lending with agricultural activities. It is interesting to recall that the Central Banking Enquiry Committee (1929) [Chandavarkar, 1982] suggested that partnership arrangements should be developed along the lines of the 'Kommandit' practice in Germany, whereby a bank, instead of opening a new branch, became the financial partner of a local moneylender. 'This would have the advantage of combining local knowledge and unlimited liability of the money lender without incurring heavy overhead charges on a new branch'.
- The banks should provide opportunity to their rural clientele to meet the designated grievance redressal officer in all bank branches without prior appointment during second half of the working hours in fixed working day.
- The banks should strive to deal with their clientele in an open and transparent manner. They also should explain the rationale behind bank's decisions regarding loan applications, consistent with requirements relating to business principles.

In the ultimate analysis it can be concluded that both demand and supply constraints are responsible for low credit coverage and hence low productivity in the agricultural sector. The commercial value of agriculture, particularly in Cachar district is quite absent, because it is working under a regime of many negative factors, such as, low production, indifference in quality standards, fragmented land holdings, and the backward technology and poor infrastructure, unreliable credit supply, politicization, attitude of non-payment of loans, the indifference of government officials who are primarily responsible for implementation of schemes as well as their failure in carrying out their entrusted responsibilities for building up ability of farmers for repayment of loan etc. All these factors are producing many conflicting and degenerating factors in the process of credit delivery in the agricultural sector by the commercial banks.

In present scenario of agricultural economy in Cachar district, participation of a properly combined group of caretakers based on local needs may be tried with benevolent as well as commercial attitude for upgradation of the farmers assuring repayment of loans by increasing recycling of loans, where the zeal in the mind of farmers and lenders will be visible. A modernized agricultural sector which is not starved of credit may thereby play a vital role in expanding the process of economic reforms.

NOTES

- The District of Cachar located in the Southern most part of Assam is one of the oldest districts of Assam. Cachar became a part of Assam in the form of a district on 6th February, 1874 and Silchar became the Sadar subdivision. North Cachar Hills subdivision was formed in 1880. Since independence, Cachar district had four subdivisions, viz. Silchar, Hailakandi, Karimganj, and Haflong.

  Haflong subdivision was later on separated from Cachar and subsequently it was constituted into the present district of North Cachar Hills. The subdivision of Karimganj was upgraded to a district on 15th July, 1983 and Hailakandi subdivision was also upgraded to a district on 1st October 1989 vide Govt. of Assam's Notification No. AAA. 98/89/Pt/ 1, dtd. 29/09/89. Lakhipur subdivision was formed in June

1988, vide Govt. Notification No. AGA. 544/86/118, dtd. 20th June, 1988. At present Cachar district has two subdivisions - Silchar and Lakhipur. Cachar district alone accounts for 54.7% of the total area of the Barak Valley. The district covers an area of 3786 sq. kms. The district is located between $24^0.22'$ and $25^0.8'$ North Latitude and $92^024'$ and $93^015'$ East Longitude. The District is very strategically located as the supply and communication line to the Mizoram, Tripura and Manipur passes through it.

- Government of Assam in the year 1964 requisitioned about ten thousand bighas of evacuee land in Cachar district for settlement of migrants from erstwhile East Pakistan who came over to India as 'refugees'. These lands spread over a large area in Silchar agricultural subdivision. Relief and rehabilitation department had given them settlement on the evacuee land and other khas land. These refugees as mentioned above were not illegal occupiers. They were given allotment of land some 35-40 years ago and during this long period they made these lands usable by their manual labour and also investing their small money in almost all cases. It is clear, these refugee people are not to be treated as 'floating population'. Since the days of allotment these poor people are continuously residing there and as such their names are also entered in the electoral roll as per rules under Nehru-Liyaqat Pact. They are allottees as per govt. orders and they are in the land for about 40 years or so. They are not in illegal possession of the lands they hold. Simply govt. is to regularise their cases. Bank may ask them to apprise them of the orders of allotment for their total satisfaction in the matter of issuing agricultural loan etc. (*Source:* Gathered from Revenue Department officials who worked previously and field visits in the sample villages of Cachar).

1. *Touzi Bahir Revenue*: The term 'Touzi' means 'patta'. The land which is outside the patta land is known as Touzi Bahir land. 'Touzi-Bahir Revenue' is a fine imposed by the Govt. on illegal occupants of the khas land of the government. If for three consecutive years the occupants pay Touzi Bahir Revenue on the alloted land then they are given 'Eksona Patta'

on application by the government. If this system continues for some more years they may be considered for giving 'Meyadi Patta'.

2. *Iqrarnama*: It is an Arabic word which means 'affirmation', 'acknowledgement'. The landholders of cultivable lands by registered document virtually sell the land at much lower rate. Certain quantity of lands against certain amount of money specifically mentioning in the document that within a specified period of time, say 5 to 10 years, money will be returned to the creditors. Otherwise, the land will remain in possession of the money lender. The time depends on understanding of the vender and vendee. The land in question in such agreement shouldn't be utilized for other than agricultural purposes. The money-lender in such cases get absolute right on the produces of the land concerned, during the period of contract. This benefit, in other sense, may be termed as 'interest'.

3. *Swaran-lipi*: This system is followed by land holders who are in need of money for meeting production expenditure as well as consumption expenditure. The agreement of transfer of land is signed by seller and purchaser in front of two-eyewitnesses. Paper measuring 13" & 8" with requisite stamps fixed thereon is used for this agreement and this is not required to be registered.

   Some people who are in possession of government ceiling surplus land, and grant areas are in practice of selling certain portion of such lands through "Swaran-lipi". In the eyes of law, such transfers of land are not valid. Also Govt. policy of providing land to the landless people is not always taken place in reality.

   A separate document (called assurance letter) is made by the buyer assuring the seller that if money is returned within the stipulated period of time, then the buyer will return his land. This assurance letter should be registered. It is worth mentioning that land cannot be forcibly converted to ownership land by Iqrarnama within due time. In case the debtor-farmer fails to return the money in time or within the extended period of time, the land stands vested in the name of money lender and consequently the loanee-farmer will be deprived of his land which was given at a comparatively lower price than market rate. Transaction as per Iqrarnama involves considerably big amount generally varying from Rs. 10000 to 1 lakh as was found among the sample households. Iqrarnama relates to agricultural land only.

4. *Izara*: Izara are of two types (a) Verbal Izara and (b) Written Izara.

   It depends on mutual faith, the govt gets no revenue. Non-judicial stamp or on white paper measuring 13" x 8" written and signed by moneylenders and sellers in presence of two witnesses. It has no legal sanction.

   Generally 'IZARA' are required for small amount of money, say about for 1000/- to 5000/- against cultivable lands measuring 1 to 2 bighas. Such agreements under 'Izara' are made for a specified period of time. If the debtor cultivator fails to return the loan within the specified period, the moneylender forces the farmer to arrange for refund by selling the land in question or other property. Interestingly, some lenders are seen interested to get benefits from the production of the land mortgaged to them. They do not press the farmer for quick return of their money, as I gathered during my field survey.

   'Izara' is the most popular and widely prevalent system of taking money against land mortgage nowadays in Cachar. This system has no legal sanction.

REFERENCES

Basu, Santonu (1999), 'Why Institutional Credit Agencies are Reluctant to Lend to the Rural Poor: A Theoretical Analysis of the Indian Rural Credit Market', *National Bank News Review*, Vol. 15, No. 1 NABARD, Mumbai.

Bhattacharjee, S. (1984), 'Eastern India' in Dharma Kumar and Tapan Roy Choudhury (eds.), *The Cambridge Economic History of India*, Vol. 2, Orient Longman Ltd.

Blackman, Allen (2001), 'Why Don't Lenders Finance High Return Technological Change in Developing Country Agriculture?',

American Journal of Agricultural Economics, American Agricultural Economics Association, Ames Lowa, Vol. 83, No. 4.

Bouman, Frits J.A (1989), *Small, Short and Unsecured Informal Rural Finance in India*, Oxford University Press.

Dastidar, P.K. Ghosh (2000), 'Rural Banking in India in the New Millennium - Need for Paradigm shifts', *National Bank News Review*, April-June 2000, Vol. 16, No. 2, NABARD.

Dey, Nikhil Bhusan and Rakshit, Sudipa (2003), 'Problems of Agricultural Financing by Commercial Banks in Cachar District of Assam - An Overview', in K. Sengupta and N. Roy (eds), *Economic Reform and Agricultural Development in North-East India*, Mittal Publications, New Delhi.

Ghosal, S.N. (1972), *Agricultural Financing in India*, Asia Publishing House, Bombay.

Kalam, A P J Abdul and Rajan, Y S (2002), *INDIA 2020, A Vision for the New Millennium*, Penguin Books, New Delhi.

Lipton, Michael (1989), *New Seeds and Poor People*, Heritage Publishers.

Majumdar, N.A. (1996), *Financial Sector Reforms: An Exercise in Introspection*, March 23.

Mohon, Rakesh (2002), 'A Decade After 1991: New Challenges Face in the Indian Economy', *RBI Bulletin*, Vol. LVI, No. 11.

Rakshit, Sudipa (2006), "Performance of Commercial Banks in a Agricultural Financing : A Case Study of Cachar District, Assam during 1990-01 to 2003-04" (unpublished Ph D thesis).

Rudra, Ashok (1992), *Political Economy of Indian Agriculture*, K.P. Bagchi and Company, Calcutta.

Stiglizt, Joseph E. (1994), 'The Role of the State in Financial Markets', *Proceedings of the World Bank Annual Conference on Development Economics, 1993,* Washington, D.C.: World Bank, 1994.

Reports

Agricultural Status Paper, 2002-03, District Agriculture Office, Silchar, Cachar.

Annual Credit Plan (1990-91 to 2004-05), Lead Bank Office, UBI, Silchar.

Annual Credit Plan (2001-2002, 2003-2004).

Assam Human Development Report 2003, Government of Assam.

Basic Statistics of North Eastern Region 1991, NEC, Shillong.

Bhattacharjee, Shekhar (2001), *District Profile*, Department of Agriculture, Govt. of India, Cachar, Assam

Cachar Gramin Bank, *Annual Report 1992-93, 1995-96, 1996-97, 1998-99,2003-04,* Silchar, Cachar, Assam.

Economic Survey of Assam (1990-91 to 2004-05), Directorate of Economics and Statistics, Assam, Guwahati.

Economic Survey of India (1990-91 to 2007-08), Govt of India, New Delhi.

RBI (2005), Agricultural Credit in India: Status, Issues and Future Agenda (Rakesh Mohan), *RBI Bulletin.*

Tenth Five Year Plan (2002-07), Volume I, Planning Commission, Govt of India, New Delhi.

Tenth Five Year Plan (2002-07), Volume II, Sectoral Policies and Programmes, Planning Commission, Govt of India, New Delhi.

Potential Link Credit Plan (2003-06 to 2008-09), Cachar District, NABARD.

The World Bank in India (Nov. 2006), Vol. 5, No. 3, p.2, The World Bank.

Seshadri, R.S. (2008) in *Kamal Sandesh,* Vol. 3, No. 8, April 16-30.

8

Locating Women in Economic Participation: The Case of Barak Valley

—*Ruma Paul*

The process of development and expansion of the global market have witnessed an increase in the participation of women and through the course of various studies, it is now evident that women is identified as distinctly different from men not only in terms of biological differences but also in essence and understanding. However even with the process of modernisation sweeping in. the role and function of women is governed by patriarchal powers and, the very construction of women is gendered and is identified as being subservient to men. The 'sexual division of labour' is only a reasonable understanding and compatible arrangement between consenting adults – men as the provider, and women the nurturer, no force is involved. [Banerjee, 1995:53-69]. However, this segregation of roles on the basis of gender has been not only with respect to the social construction of women, but is also explicit in various kinds of work that both men and women perform. A comparison with the nature of work of both the sexes' reveals that women have not been able to catch up with men in terms of employment, share of wages, time devoted to work; in spite of their efficiency, productivity, and capability. Women have been denied equal footing with men in terms of equal pay, remuneration, decision making authority and status. Exclusion of women from these rights only retard the progress and development of nation because it is, by now, universally accepted that women workers constitute a meaningful category and their participation along with men constitute the total labour force in the economy.

Experience of developed as well as developing nations reveal that fruits of economic growth have not only failed to trickle down to men and women alike, but has instead further widened the gender gap. In the economic sphere, there is gross under valuation of women's contribution and also absence of recognition of their contribution. Women in every nation work for considerably longer hours than men folk, but major portion of their work remain unpaid or under paid in contrast to their male counterparts. Since the income earned by an individual is an important determinant of a person's economic status, a woman deprived of these legitimate rights often suffers from low economic status, which in turn gives rise to low sociological, political, and psychological status in the society.

Significance of women's economic participation

Participation here means the contribution made by the women by taking an active part in the economic activity of a nation. Generally, this contribution is measured in monetary terms so that it can contribute to the GDP of a nation. The female work participation rate determines the percentage of women working which necessarily helps to indicate whether they are fully employed drawing wages or marginally employed with minimum wage or remains under paid. Workforce participation occupies an important place for investigation because it interacts with various demographic variables. For both the sexes, there exist a strong casual relationship between occupation and work, fertility, health, income, literacy, migration and mortality [Fong, 1975]. There is close association between all these factors and the participation rate of women. Therefore, there is a need for studying FWPR in relation to these variables because a woman's entry, existence and increased participation in the labour market are related to these factors.

Situation in India

In India, the patriarchal ideologies have always governed women in socializing and construction of women as a highly flexible resource of the household. According to Banerjee [1995 : 53-69], such patriarchal ideology dictates the functioning of woman as an essential component of her socializing process. The ideologies such as chastity, universal early marriage, and frequent motherhood prevent women from entering the labour market and in situations where they do, it is by compulsive economic condition like poverty and deprivation. While the economic necessity and the need for supplementing family income could be the major factors for women to enter the labour market, economic transformation, industrial progress and technological advancement could also be cited to constitute an equally important reason [Usha, 1981]. Women are gradually moving from the traditional sectors to that of paid work. This has increased the participation of women workforce in the labour market. It is by now unanimously accepted that women play an important role in the economic development of a nation. The development of an economy can not only be measured in terms of GDP, per capita income and wage but it needs to be seen also in terms of effective participation and utilization of the capabilities of both men and women as an effective human resource potential. In this context, the economic role of women cannot be isolated from the total framework of development. Equal participation of women is a pre-condition for the development of not only the women but also the country as a whole. Emancipation of women, their equality with men are impossible as long as women are excluded from economic productive work. The existence of the concept of inequality of the sexes requires that in the present context we rescue the notion of gender from its ritualistic incantations and make it effectively work for a more emancipatory and inclusive social order [John, 1996 : 3077].

Existence of discrimination

Research studies have demonstrated a universal fact that there exists a wide gap in the earnings of men and women though the technique, methodologies and sample data have been different, yet it is an unanimous agreement among all studies that wage discrimination does exist. Although overt wage discrimination is illegal yet most employers discriminate by segregating jobs or by offering less training and skill to economic workers and justifying their low wages. The I.L.O. report (1943) states that

> partly on account of thc long standing traditions, social customs and preconceived notion that women are inferior to man in respect to their average outturn, and partly due to their shorter experience and inferior vocational training which in turn, lessens the demand for female labour when demand and supply have free play in the employment market. (Sic)

Denial of rights has also led women to work for low wages, to be dependent on the male bread earner and to be dependent earners rather to be an independent or self dependent worker. This as feminist economists argue are the limitations chained to women within the walls of patriarchal ideology which establishes that through such denial process, women are self convinced to work as marginal and dependent workers [Agarwal, 1995].

Productivity of women to generate income is closely related to the amount of space allocated to her to operate independently as an economic agent. In majority of cases this allocated space is very much restricted as women are considered to be a flexible resource in the work participation process and investment in women in the form of training, education for any particular skill or field of operation is considered to be either wasteful or is ignored. This widespread discrimination has led feminist concern to claim for equal pay for man and women workers.

Gender studies and feminist critics point out that the employment of women is largely governed by the stage of economic development of the society. In an agriculture dominated economy, e.g., it is found that while women and men jointly work together in the family farm but with improvement in the living conditions and a rise in the standard of living, it is the women, who by societal prerequisite opt out of work and thus their participation in economic activity declines adding to it a sense of deprivation and a loss of income [Mitra, 1997]. In agriculture, the impact of modern technology and shift from subsistence to cash cropping have largely benefited men, while increasing women's work within the family and leading to a rise in the number of casual labourers [Boserup, 1970 : 117]. While the new opportunities of employment are potentially for empowering women, the forces of economic volatility, job insecurity and loss of livelihood which have been inimical to gender equality have also been unleashed [*The Times of India*, 23rd Oct.1999].

Structural technology and state policies have not been able to deal with these issues in a constructive manner. Capitalism through structural reforms has used new technologies to exploit gender differences in the labour market through persistent segregation as secondary (unskilled) workers with relatively low income and substituted women for higher paid men. The long-term effect of such a change is low operational cost and displacement of women through feminization of women work force [Banerjee, 1999 : 299-317].

In India, discrimination in the labour force exists in various forms: discriminatory wages, occupational segregation, human capital discrimination and so on. While differentials in male-female unemployment rate reflect employment discrimination and pre-entry human capital discrimination, wage differentials that exist are indicative of discrimination at the work place.

Brief Socioeconomic Scenario of the Area under Study—Barak Valley

Barak Valley comprises the three districts of Cachar, Hailakandi and Karimganj lying in the southern part of Assam with an area of nearly 6922 square kms. The economic picture of Barak Valley greatly changed after the British annexation. This has been mainly with respect to tea industry, cottage industries and agriculture [Chatterjee, 2000: 145]. Annexation helped the people to liberate themselves from their previous confinement and gradually move towards modern lifestyle. After the creation of the Silchar Town in the year 1834 A.D., there was a gradual development of urban livelihood with appointment of various people both men and women in different official posts. The expansion of tea-plantations in various parts of Cachar led to the development of new roads, schools and health centres. This was also partly responsible for the influx of labourers from outside the valley [Biswas, 1998: 1-5]. Though agriculture has always been the main source of livelihood of the people with the fertile lower soil covered with rich crops of mainly rice, the high lands within the valley have been mostly utilized for tea plantation [*Statistical Profile of Barak Valley, 1998]*. The introduction of the tea industry led to the establishment and creation of employment opportunities of both men and women in paid employment. This has also resulted in large-scale influx of labourers from outside the state where women have entered the work force leading to supplementary earnings for the household [Dutta Ray, 1978: 192].

Over the years, in Barak Valley, there has been a rapid rise in the total population as well as female population (annual growth rate of population was 1.89% and 1.93% respectively during 1991 census with reference to 1971 census). Sex-Ratio which indicates the number of females per 1000 males, have also increased for all the three districts except the district of Karimganj which has registered a very nominal fall. This indicates a healthy trend with no such social taboos and discrimination attached to a girl child and which is also reflective of a positive attitude towards the girl child, which may have developed through an increase in literacy rate in the state. This has however not changed the relative perception towards women in providing equal opportunities in related fields of agriculture and other services. The increase in the proportion of female workers in agriculture and allied activities from 59.6% in 1961 to 77.5% in 1991 census shows that agriculture still occupies a predominant sector of livelihood. Majority of these workers are engaged as agricultural labourers and few are engaged as cultivators signifying that majority of them work as casual workers with no independent decision making ability and no control over their land.

Report of the employment exchange though provides a hopeful picture of increasing awareness of the people seeking for employment – however does not provide respite in the nature of providing employment opportunities to the growing number of literates and educated turnouts of both male and female population. This also indicates that with the diminishing absorption of people in the government services, the dependency of government jobs is gradually diminishing and this is reflected in the low rate of registration in the employment exchange offices. And even when it does exist; males are always preferred to females in the service sectors.

Apart from the engagement of women in various occupations, women of Barak Valley are also engaged in various household activities, which generate considerable income to the respective households. The folk workmanship of Barak Valley could be divided into two groups, one, that has still retained its traditional structure using age old techniques of craftsmanship and another that has lately introduced modern tools and implements within the household sector. The household sector is dominated by women who participate in various work involving sola craft, bamboo handicrafts, kantha embroidery of intricate designs. In addition to spinning and weaving within the household, women assist the men-folk in repairing of houses, raising crops like weeding, transplanting, reaping, thrashing and winnowing crops. Veiling or purdah system never existed and therefore the mobility of the women was not restricted in the Valley [Sarma, 1989 : 123]. Duttapur village of Karimganj was once famous for fine embroidery work where women were traditionally bound not to receive any remuneration for their efforts and were distributed or utilized for use within the family. Therefore marketing of these products or adding a monetary value was never practiced. Thus productivity of the worker was never valued. In recent years this art form has received commercial status and as a result many women have been benefited in terms of paid work as well as an enhancement in their status [Basu, 2000: 214-220].

Among the other social communities, the Manipuri women constitute to be very hard working and enterprising lot of women. The Manipuri women have retained an almost monopoly over their traditional silks of weaving and colourful handloom products [Okram, 1994 : 149-53]. Among the Dimachas of the plain area, women specially belonging to the Barman community are very hard working and industrious and are less dependent on the male members and enjoy economic freedom [Chatterjee, 2000].

Though fish trade had always been an important form of economic activity in Barak Valley but women engaged directly or indirectly in this trade have been almost negligible. It is to be noted here that certain foothills of the plains of Cachar, Hailakandi and Karimganj are covered by forests. They were reserved forests and the law prohibited the human settlement in such forests [Chatterjee, 2000 : 34]. Therefore, these forest resources also constituted an important source of livelihood for most tribals and plain people of Barak Valley. But these sectors have not yet developed as an organized form of employment generating sectors, which could gainfully provide employment opportunities to the women folk. But an understanding of these works will certainly reveal that most of these works are small-time works confined within the household and are yet to find sufficient exposure in the market. Overall the macro perspective suggests that there are ample opportunities to find alternative income sources for the women workers especially in the rural sectors of the economy in the form of self-employment schemes and upgradation of traditional crafts.

In spite of various forms of activities, women of Barak Valley are yet to see the face of the true form of women emancipation and development. The economy suffers from various infrastructural and development deficiencies. Unemployment is rampant in the Valley with no simultaneous development or growth of job-generating sectors. Though this concerns both men and women alike, but the problems faced by women are more acute. Therefore, striving towards equality between the sexes is a much 'market compatible approach' and also in consonance with prevalent macro economic policies [Sudarshan, 2001 : 21]. Thus, women need to be viewed and empowered on an equal footing with men in all spheres of life.

Methodology

Study is an attempt to locate the structural pattern of

participation in economic activities by the women of Barak Valley and relate it to some social economic variables. This has been considered necessary to assess the impact of such variables of women's economic participation. The result of sample size of 645 populations cutting across rural, urban, linguistic and religious groups is briefly analysed.

Participation in an Unequal World and an Analysis of Secondary Data

Women in Assam constitute almost 48% of the total population with almost 21.61% of the female population working according to 1991 census. Majority of the population are engaged in the informal sector and therefore are subjected to the discrepancies, which exists in the labour market. Traditional handicrafts, like weaving and bamboo works occupy a prominent place next to agriculture. In Barak Valley the impact of the developmental activities on women during the post-independence period is significant. The role and status of women in Barak Valley have undergone notable changes with improvement in literacy and awareness that comes with development and change in the social and economic life. Yet inequality of status persists in different aspects of economic, social and family life. Inadequate employment opportunities compel women to concentrate in the unorganised sector and women lag behind men in employment in salaried jobs and in sectors, which involves decision-making and exercising power.

The worker population ratio presents a broad picture of the trend of female work participation vis-à-vis male and total work participation rate in India as a whole and with urban rural division. This is based on census and NSS reports. (ref: Fig. 1 & Table 1) The picture that emerges from the table, displays the following facts:

- The Female Work Participation Rate in India is far

behind that of males for the entire period of our analysis irrespective of the definition of workers.

- The participation rate for both female and male is increasing over the years.
- As regards the 1971 and 1981 censuses there was a considerable concern about the undercount of female workers. And hence 1991 census was redefined to include 'unpaid workers'. In spite of these changes FWPR according to 1991 census showed only a marginal increase.

Female population constitute a strong force in Assam and the proportion of female workers in work participation shows an improving trend like the rest of India but reveals a declining trend during 1971 which could be attributed to the definitional change of 'workers'. The latest census reveals that FWPR in Assam is comparable to FWPR in India. The FWPR, according to 1991 census is 21.61% as against 22.05% in India for the same period. Correspondingly male WPR in Assam during 1991 census was 49.44% as against 51.08 % in India.

The total FWPR of Barak Valley of South Assam although lags far behind India and Assam, has also shown an improving trend outpacing the rate of growth of female population (Table 2). The percentage of total workers has shown a steady rise, with the district of Cachar registering the highest FWPR. A comparison of FWPR of the 1991 census with the 1961 census shows that in percentage terms, the FWPR has improved from a negative growth rate to a positive growth (Table 3). FWPR has significantly increased in the valley from 4.93% in 1971 to 11.38% in 1991 with the largest concentration of the Female Work Participation Rate in Cachar district (13.08%) followed by Hailakandi (11.97%) and Karimganj (8.6%) during 1991. The growth rate of FWPR has widely varied among the districts varying between 3.90%

(minimum) for Cachar, 5.52% (maximum) for Hailakandi and 4.29% in Karimganj in between according to the 1991 census. However, except this recent rise in FWPR, which as explained earlier, is perhaps due to definitional change. The overall growth rate of WPR has lagged behind the rising growth of population throughout the period since 1951. In the overall analysis the following facts emerged:

- The FWPR is much lower than the MWPR in India and the same trend follows also in the state of Assam and Barak Valley as a whole.
- The overall FWPR is always found to be better in India and Assam compared to Barak Valley and its three districts, viz., Cachar, Karimganj and Hailakandi.
- The annual growth rate of FWPR during the census of 1991 with respect to earlier census is found to be much better, both in India and Assam compared to Barak Valley and its three districts indicating that the position of women in entire Barak Valley needs to be improved in the labour market and also a large section of the women population can be actively involved in various productive activities.

Table 1

India's worker population (ratios by sex and rural / urban residence)

Sl. No.	*Year / source of NSS round*	*India*			*Rural India*			*Urban India*		
		P	*M*	*F*	*P*	*M*	*F*	*P*	*M*	*F*
01	1951 CENSUS	39.1	53.9	23.4	39.5	53.5	25.0	37.1	56.4	14.7
02	1961 CENSUS	43.0	57.1	28.0	45.1	58.2	31.4	33.5	52.4	11.1
03	1971 CENSUS	34.0	52.7	13.9	36.1	53.6	15.5	29.6	48.9	7.1
04	1981 CENSUS	*36.8	52.6	19.8	38.9	53.8	23.2	30.0	49.1	8.3
05	1991CENSUS	**37.5	51.1	22.1	40.0	52.5	26.7	30.2	48.9	9.2

* Excludes Assam, ** Excludes Jammu & Kashmir

Source : Censuses of Population / NSSO Rounds, Govt. of India
P : Total Population, M : Male, F : Female
Source : (i) NSS (1990), Government of India, New Delhi; (ii) Census of India 1961, 1971, 1981, 1991, (iii) Papola, T. and Sharma, A (1999) : *Gender and Employment in India*, Vikas Publishing House Pvt. Ltd., New Delhi, pp. 26-27.

Population and Economic Participation

A study for the period 1951 and 1991 reveals that overall population has increased throughout with India maintaining lead followed by Assam whereas Barak Valley lagging mostly behind. This is indicated in Appendix 3 (Table 4). The yearly growth rate of total population increase for India between 1951 and 1961 has been 1.98% increasing to 2.24% between 1961 and 1971 and further to 2.2 % between 1971 and 1991.

In Barak Valley while the growth of population has registered to be 1.89% but the WPR growth has risen by a meagre 0.34%. This clearly shows that overall growth rate in WPR have never kept pace with the overall growth rate in population. Similarly, the growth rate of participation of females shows a varying trend while growth rate of population was steady. However, a study of 1991 census with reference to 1971 census has shown a significant improvement of female participation as compared to MWPR and overall WPR. While this may present a very optimistic picture, however, the participation details of female workers show that in the overall situation, along with the growth of population and rise of WPR, female participation has remained below the MWPR and total WPR. Further FWPR has not improved significantly and women are always discriminated when the question of providing stable and assured jobs arises. However, the sudden spurt in the rate of growth of FWPR during 1991 with respect to 1971 is perhaps due to change in definitional aspect in census and actually no significant improvement in women's status has accrued. This is further confirmed by the fact that FWPR during 1991 has remained below the FWPR of 1961 and a sudden drastic unexplained drop preceded the increase in growth rate of FWPR during 1991 during 1971 with respect to 1961 census. From 1971 to 1991, the growth rate of FWPR and female population for India were 8.33% and 2.19% respectively whereas for the same period Assam and Barak Valley have shown a growth rate in respect of FWPR (7.97% and 4.27%) as against the population growth rate of 2.24 % and 1.93 % respectively. The latest census study also reveals that in respect of FWPR, Barak Valley is almost half to that of Assam and India. In Barak Valley though the overall population has been increasing it has not been commensurate with a simultaneous expansion of service and other employment creating sectors. Sex ratio and literacy rate, which works out to be fairly high in the valley, is indicative of a growing increase in social awareness among the people. However non-absorption of the growing number of jobseekers has failed to harness the full potentiality of women as an important economic category of population.

The female sex ratio in India has always been found to be adverse to women and has been declining over the decades with slight improvement during 1981 census. Assam, however, shows an increasing trend throughout the entire period. The same increasing trend is also found to be in all the districts of Cachar. The steady rise in sex ratio and literacy rate in Assam and Barak Valley could be an indicator of the increase in social awareness amongst the people. This, coupled with an improvement in the FWPR could harness the full potentiality of women as an important economic category of population.

It is thus clear that with the overall increase in population, FWPR either in Barak Valley, Assam or India is much lower vis-à-vis MWPR. With the increasing population, combined with a poor representation in the total workforce, the unemployment in the female population is found to be alarming. The dependency ratio i.e., proportion of non-workers to total population is high and unemployment both in disguised form is equally rampant in the districts. The increasing working population is again heavily concentrated in rural areas than in the urban sectors. This shows that there exist symptoms of reliance and subsistence economy where marginalisation of women workforce is high and situation of being underpaid, unrecognized and un-remunerative jobs are common.

Female Economic Participation in Barak Valley - An Empirical Account

An examination of the primary data reveals that the district of Cachar registers the highest participation of women labour force among all the other districts of Barak Valley. An investigation into the villages of Barak Valley has revealed that participation of women in work are found everywhere and women are engaged in various forms of work such as weaving, brick-making, agriculture, plantation, construction, poultry farming and also as fish and vegetable vendors. According to *Assam District Gazetteer* (1991) while many works are done with the assistance of or in co-operation with the male members, there are many women who work individually on their own initiative so as to supplement their family income. Unmarried women also work under family compulsion and at many times earn so as to save their earnings for their marriage expenditure. But most of these works, which they perform, arc either not recognized nor are they substantially rewarded in monetary terms. The vulnerability of these workers acts as a source of cheap and exploited labour class [Vanamala, 2001]. In the absence of trade unions or unionisation of work, workers feel insecure and uncertain about their future job security. Typically, the female workers on account of their low level of literacy and low bargaining power are subjected to the terms and conditions laid by their employers. In other words the relation between the employer and the employee is typically a relation between the 'power possessor and the one who is not the bearer of such a power'.

The job scenario of female workers in Barak Valley is therefore not very different from the all India scenario. 'Feminization' of the informal sector has assumed a tremendous proportion. With the adoption of the market-based economy, several traditional industries have been adversely affected through competition. A large section of handloom industry, which has been the second largest employer of women labourers, had been badly hit by steep rise in the prices of cotton and cotton yarn. The sericulture industry, which occupies an important place in the economy of the region, is severely hit by the competition from imported silk cocoon. The household and cottage industries, which used to employ a substantial portion of women labourers, have also lost government patronage and thereby unable to withstand the stiff competition from low priced better-finished industrial products. Decline of such activities, where women labour force had played a dominant role implies that they are forced to combine several economic activities in the unorganized sector, which are extremely low paid and without any job security in an attempt to supplement their family income.

Labour market segmentation on gender basis

Table 1.1 documents that in the entire Barak Valley and its three constituent districts; FWPR is much lower than the overall MWPR. While FWPR is worked out to be 22.45%, MWPR is estimated at 57.27%. The same trend of low participation of females has been observed for all the districts of Barak Valley with Hailakandi registering the lowest female participation rate (19.64%). It has been observed that in the rural areas of Barak Valley, both male and female workers are engaged in agriculture related activities and in Barak Valley alone, almost 13% are engaged in agriculture related activities in the unorganised sector and the largest concentration is found in Karimganj district (17.7%).

The low participation of females in Barak Valley could be partly explained due to the factors like lack of education and skill and traditional societal set-up in the villages and prejudices. Though female WPR in Hailakandi is low but when distributed over rural urban population it is observed that Hailakandi has a larger proportion of urban female work participation rates (21.65%). But in the other two districts the scenario is quite different. This could be due to the fact that overall job opportunities are limited in Hailakandi and even in the rural areas the employment situation is very grim and, therefore, whatever job opportunities that does exist it is in the urban areas. Moreover, two towns have been considered to represent the urban area in Hailakandi, one is Hailakandi town by itself and the other Panchgram. Of the entire valley both these towns represented a better scenario with respect to urban participation rate both for males and females. Karimganj and Cachar follow the general trend where lack of opportunities in the organized sector, which requires greater skill and formal training, restricts the absorption of females in this sector which is assured of stable wages and secured job. This leaves them with no alternative but to crowd in the unorganized sector or subsistence sector both for the females as well as the males. The highest female as well as male participation in Cachar is due to its logistic advantageous situation. Cachar records a high female participation of 23.52% and male participation of 58.90%.

The growth of WPR in the district of Karimganj may also be explained due to its closeness with the borders of Bangladesh and Tripura state. Pisciculture or fishing cultivation is widely practiced in Karimganj area. The area under pisciculture in Karimganj is 11,650 acres (*Assam District Gazetteer*, 1991) and is a popular occupation between both the sexes. Similarly candle and soap making factory also employ a sizeable proportion of workers in the district.

The growth of several cottage industries in the valley has also led to creation of overall employment opportunities in the valley. In urban areas, the socio–cultural prejudices against female employment are found to be less strong due to higher level of awareness that develops with higher literacy, industrialization and modernization and this could be one of the factors for which male-female gap is higher in the rural areas. Along religious lines, women's proportion is the lowest in Hailakandi district among the Hindus, whereas Muslim women record the lowest participation both in Karimganj as well as in Hailakandi. Among the linguistic groups too, work participation is low both among the Bengalis, who are the dominant linguistic group of the region as well as the non-Bengalis too. Therefore cutting across various districts, social, linguistic and religious groups, labour market segmentation is more acute in Hailakandi district.

The incidence of non-working females is substantially higher in urban areas than rural areas in all the three districts. In urban areas of Cachar and Karimganj districts almost 78.86% and 79.44% respectively of females in the productive age group (15-59) are unemployed. Both these are higher when compared to the rural unemployment (Cachar 76.08% and Karimganj 77.72%). This may be due to migration of jobseekers from rural to urban areas. On the contrary, in Hailakandi district the unemployed females are concentrated more in the rural areas (81.89%) whereas in the urban areas it is comparatively less (78.35%). Job opportunities in the urban areas of Hailakandi are so very limited thereby increasing the pressure in the rural areas. Migration, which has taken place in the district, has taken place directly from the rural areas. The reasons for this relatively low unemployment trend lies in the fact that, most of the rural women who do not work directly for any monetary gain are engaged through out the day in various household activities like child rearing cooking etc which are not counted as economic activities and thus do not enter in any statistical accounts.

Table 1.1
Work Participation in different categories

District Valley/	*WPR*	*For valley/ dist.*	*Work Participation on the basis of*							
			Rural/Urban		*Religion*		*Language*		*Caste*	
			Urban	*Rural*	*Hindu*	*Muslim*	*Bengali*	*Non-Bengali*	*SC*	*ST*
Barak Valley	M	57.27	54.15	57.88	58.97	54.02	56.08	60.25	58.01	57.88
	F	22.45	21.10	22.81	24.18	19.56	22.68	21.77	24.01	22.09
Cachar	M	58.90	56.00	59.23	61.09	54.78	57.30	61.65	60.23	58.06
Dist	F	23.52	21.14	23.92	24.95	22.41	23.53	23.49	27.16	22.13
Karimganj	M	57.58	53.43	58.53	58.08	56.57	56.91	61.02	56.47	55.56
	F	21.91	20.56	22.28	24.16	15.38	22.71	17.72	21.43	22.22
Hailakandi	M	52.98	53.33	52.83	54.80	50.42	52.59	54.29	54.29	50.00
	F	19.64	21.65	18.11	21.13	15.79	20.11	17.14	17.54	20.00

WPR: Work Participation Rate, M: Male, F: Female
Source: Field Survey.

Participation in the organized and the unorganized sectors

Employment outside the defined organized sector is employment in the unorganized sector or the informal sector including self-employment where the relationship between the employer and the employee is informal. An unorganized sector includes all unincorporated enterprises and household industries other than the organized sector units in the private sector. Unorganized sector accounts for about 60% of the total net value added in the economy [Economic Intelligence Service, 1998].

Table 1.2
Job Distribution of Female Workers (in percentage)

Sl. No.	*For the Valley / district*	*In Organized Sector*	*In un-organized Sectors*			
			Household	*Private*	*Agriculture*	*Others*
01	Barak Valley	23.7	13	19.8	29	14.5
02	Cachar District	22.7	10.6	17.7	29.1	19.9
03	Karimganj District	24.1	17.7	27.8	24.1	6.
04	Hailakandi District	26.2	11.9	11.9	38.1	11.9

(*Source*: derived from Paul. R. "Structure of women's participation rate and its determinants in Assam with special reference to Barak Valley." Ph.D. thesis (unpublished), Assam University, Silchar, 2002).

Table 1.2 shows the distribution of female workers and the unorganized sector in Barak Valley and the three districts. In Barak Valley almost 76.3% of the total female workers surveyed are in the unorganized sector and only 23.7% are employed in the organized sector. Concentration of employment in the Cachar district in the unorganized sector is equally high reaching to about 77.3% followed by 75.9% in Karimganj and 73.8% in Hailakandi. Therefore, all the districts in the valley are dominated by a high proportion of female employment in the unorganized sector, revealing a wide gap of female employment between the two sectors. An inspection from the field survey into women's participation in different categories of job have revealed that the largest concentration of workers are in agriculture where almost 34.4% of female workers are employed and this trend is observed for all the districts. This is also supported from census study observation where the growth of dependency of workers has been increasing in the primary sector. The total percentage of both the sexes dependant on primary sector is 71% in Barak Valley as against 74% in Assam and 67.3% in India (1991). The primary sector is the sector, which is characterized by agriculture related occupation, which is seasonal, and wage rates fluctuating. A large influx on this sector signifies that a considerable proportion of female population is engaged in low paid jobs and are most often underemployed leading them to survive under acute financial crisis.

In the absence of sufficient employment opportunities a sizeable proportion of population has taken up even household jobs. In Barak Valley, a considerable size of women is engaged in household work. Almost 13% of women are engaged in various types of household activities like attending to poultry birds, caring and feeding herds of cattle tending to home–grown vegetables plots collecting fuel or making cow- dung cakes. All of these activities in a minor way contribute to supplement the family's income. Since these sales are confined to neighbourhood or local markets these activities yield poor returns and are not accounted in official statistics (Jain, 1985). The considerably large percentage of women workers in the unorganized sector also reflects the desperation of the women workers to work being less trained and less informed. Organized sector is the sector where wages and working conditions are relatively better and job security is ensured.

This sector covers enterprises, which are governed by some legislation. Around 40% of factor incomes are generated in the organized sector [Economic Intelligence Service, 1998].

Total jobs when classified over organized and unorganized sector reveals that organized sector job represents a meagre 23.7% compared with a major chunk of the working population i.e. 76.3% concentrated in the unorganized sector for the entire valley. The reason for this low concentration of females in the organized sector is perhaps due to the complex matrix of social, economic and cultural factors. District wise distribution of the female workers shows that the largest concentration of workers in this sector is found in the district of Hailakandi (26.2%) followed by Karimganj (24.1%) and Cachar (22.7%). With rural participation very low in the district of Hailakandi, majority of the workers are concentrated in the urban sectors. The coverage of the two towns viz. Hailakandi town and Panchgram in our survey have perhaps contributed to a greater existence of women workers in the organized sector. Share of women workers in the organized sector in the district of Cachar is comparatively low which can be attributed to

1) Better job availability in the unorganized sector in this district compared to the other two districts.

2) Migration of female workers from Karimganj and Hailakandi districts to Cachar district in search of jobs due to supposedly better economic prospects of the district. Jobs being limited in the organized sector, most of these women workers are compelled to take up petty jobs at a very low wage rate in the unorganized sector.

In Barak Valley in the absence of sufficient growth of industrialisation and increase of job opportunities, and heavy pressure of population women are left out of the organized network and are forced to eke out a marginal existence in the unorganised sector.

FWPR and the participation of scheduled caste and tribes in Barak Valley

In general in Barak Valley, both the tribal and scheduled caste women have been found to have a high level of participation in comparison to all India situation though it is much lower than the male participation rate. While the overall female participation rate in Barak Valley is 22.45%, SC and ST female workers account for 24.01% and 22.09% respectively (Table 1.1). Though there exists high participation among the females in certain districts such as Cachar where the participation of SC female worker is 27.16% as against the overall participation of 23.52% in the same district. Similarly in Karimganj and Hailakandi districts ST female workers account for 22.22% and 20% respectively for the two districts, which is higher than the district average of FWPR. But a survey of the nature of these women workers has revealed that their high participation is restricted to traditional works in the unorganized sectors.

A comparison with the male participation reveals that the participation of male workers is higher than the female workers In all the districts males outnumber females in the overall participation rate. But a comparison with the male participation of the district reveal that SC male participation is much less in Karimganj district and ST male participation is less in all the districts of Barak Valley. All these factors might have contributed in increasing the work participation rate among the females. Therefore societal set-up and economic compulsions play a dominant role, rather than gender discrimination in FWPR. However, it also needs to be mentioned that there is an under current of discrimination which is still in practice even among this social group, since the majority of female workers are employed in the unorganized sector.

In all the villages that have been surveyed the tribal groups of women have been found to engage themselves more in subsistence activities which include collecting fuel, planting vegetation, liquor brewing, weaving basket, fodder collection, agricultural cropping and handloom work. In general it has been observed that SC and ST tribes have more or less equal participation in all the districts except Hailakandi which records only 17.54% among the SC and 20%among the ST. One important factor leading to high participation among SC and ST is the influence of matriarchal society within its close boundaries. Our field study has revealed that most of the SCs are a migrant group of workers who have infiltrated into the valley at various years. The women who have been interviewed work mostly as plantation workers in the various tea gardens of Barak Valley or work as construction or as agricultural workers. Among the tribes many women such as the Nagas, Manipuris, Riangs and Khasis practice settled farming or work as daily wage labourers. Women equally participate with men in the agricultural field as well as in selling the products at the local market. The situation seems to be similar to that of Tripura, where they are actively involved in permanent agriculture. It is the women who work as exchange labourers in other fields.

Social and economic parameter influencing participation—FWPR and sex ratio

The male female composition of population influences not only the form and tempo of life in any community but also vitally affects labour supply through marriage and fecundity [Sinha,1979]. The study of male female ratio or sex composition is therefore of special importance to which we turn our attention to. The general pattern of an adverse sex ratio as obtained in India and Assam also hold in the case of Barak Valley. Of course, the excess of males over females is significantly less in Barak Valley and unlike a deteriorating trend sex ratio is noticed to be improving over the years. Studies have revealed that high literacy rate, particularly female literacy has a very favourable influence on sex ratio. The slightly better sex ratio of Barak Valley may thus be attributed to high female literacy rate. One of the conclusions that one would derive is possibly the improved literacy status in the valley are yielding results.

The data in Table 1.3, documents that among the districts of Barak Valley, Karimganj shows the highest level of sex ratio 972 as against the census figure of 944 (2001). This could partially be explained due to increased literacy in the district. Among the three districts Karimganj appears to be the most literate district recording a significant increase from 44.76% to 60.09 % (between 1991-2001 censuses) for females and from 64.05 % to 73.87% among the males. Female participation is also favourably high 11.97% (1991 census), next only to Cachar. Consequently we may establish a relationship between FWPR and sex ratio, which seems to be positive in this district. Among the towns of Barak Valley, Karimganj town records the highest value of sex ratio. The concentration of railway colonies in and around Badarpur town, a reflection of literate population in the area could also be a possible reason for the high sex ratio in that area. In Hailakandi, with a very low sex ratio of 941, female work participation also reports a low percentage of only 18.11% in the rural areas. This could be due to the fact that as urban concentration of workers is high, women along with male members move to the urban areas in search of better opportunities.

Table 1.3

Sex Ratio as per field data

Sl. No.	*For the Valley / district*	*Sex Raito*	*Sex Ratio on the basis of*	
			Urban	*Rural*
01	India (as per Census '91)	927	894	938
02	Assam (as per Census '91)	923	838	934
03	Barak Valley	961	994	953
04	Cachar District	963	1000	958
05	Karimganj District	972	1032	958
06	Hailakandi District	941	962	929

Source: Derived from Paul, R. "Structure of women's participation rate and its determinants in Assam with special reference to Barak Valley." Ph.D. thesis (unpublished), Assam University, Silchar, 2002.

In Cachar, an improved sex ratio in the urban areas speaks of greater awareness and comparatively improved standards of living among the urban people, which is reflected in a positive attitude towards the girl child. Female participation has increased proportionately as well, which is the highest in the valley. In urban areas, the explanation that is usually provided for increased participation is that participation in remunerative employment increases the returns in investments in girls that will lower the dowry and the cost of upbringing the girl child. Since the status of women is relatively better in the district, such may not be the case and hence the sex ratio in the district is much better in comparison to the all India figure.

However, though the positive relationship between a favourable sex ratio and FWPR holds good in all the three districts of the valley, the percentage of FWPR on the whole is unsatisfactory. Due to the absence of urban development, the absorption of all sections of labour force is extremely slow in the valley.

FWPR and Literacy

The aim towards being literate is being pursued in the valley with the sole objective to better one's chances in the labour market. The avenues of educated or white-collar employment exist not only within the state but also outside the region. Education is therefore one major way or a process of out- migration of not only the male force but also the female workforce. A look into the literacy (Table 1.4) suggests that the literacy level is considerably high in all the districts. Among all the districts, the district of Karimganj records the highest percentage of literacy (65.35%). The high literacy growth in the district is possibly due to a high growth of literacy in Karimganj town followed by a favourable growth of female literacy in other blocks. An improvement in literacy in the district is reflected in the growth of female participation rate in the district (21.91%).

The Cachar district is marginally behind Karimganj in female literacy 63.57% rate but shows a promising trend of female participation (23.52%). Again Silchar town records a higher literacy (82.35%) but FWPR works out to be comparatively lower than Karimganj town (20.56%). It therefore transpires that in labour market, the gender discrimination is higher in Silchar in comparison to Karimganj Town and its district or Cachar district as a whole. This may be partially explained by the fact that due to job opportunities available outside the region the process of out-migration occurs. Sonai block in Cachar district shows a low participation rate (13.95%) as well as a comparatively low level of literacy (52.38%). This may be due to the fact that, majority of women are Muslim women who find it difficult in working outside the household sector due to social restrictions and social taboos which prevent women to avail of paid jobs.

Among the religious groups, it is observed that Hindu females outnumber Muslim women in terms of literacy. In both the districts of Karimganj and Hailakandi Muslim women are at par with Hindu women in literacy. Increasing awareness and need for employment opportunities as well as to get a fair deal in the marriage market, women are given the opportunity to obtain if not more, minimum level of education because the number of graduates in Barak Valley according to our survey are 29.4% among Muslims in the organized sector as against 34.9% among the Hindus. Educational factors associated with increase in women's labour force participation are of fundamental importance. In every country there is evidence that both male and particularly female participation rates increase considerably with education. Nevertheless, differences in other determinants of female participation such as social, cultural and religious factors also affect to such an extent that countries with a high average educational level for women do not necessarily have a high rate of female labour force

Table 1.4
Literacy among various groups

Sl. No.	Valley/ District	Sex	For valley/ dist.	Literacy in percentage on the basis of							
				Urban/Rural		Religion		Language		Caste	
				Urban	Rural	Hindu	Muslim	Bengali	Non-	SC	ST
1	Barak	M	65.74	80.20	62.73	66.39	65.55	67.66	60.79	53.77	67.23
		F	63.92	73.93	61.54	65.96	61.91	66.32	57.42	48.88	65.20
2	Cachar	M	66.04	81.00	64.18	66.79	66.46	68.74	60.84	57.96	69.50
		F	63.57	86.17	63.70	67.04	60.43	64.54	61.51	52.63	72.73
3	Karimganj	M	66.31	87.65	61.78	66.27	66.67	67.91	59.55	50.00	61.25
		F	65.35	79.57	61.71	65.79	66.06	69.67	47.78	41.84	50.62
4	Hailakandi	M	64.27	70.73	60.92	65.63	62.31	64.98	61.90	46.91	76.47
		F	62.94	63.36	62.68	63.51	62.18	65.78	53.25	46.67	78.57

M: Male, F: Female
(*Source*: derived from Paul. R. "Structure of women's participation rate and its determinants in Assam with special reference to Barak Valley." Ph.D. thesis (unpublished), Assam University, Silchar, 2002).

participation [Paukert, 1984]. Among the three districts, Hailakandi shows the lowest level of literacy (62.94%) along with low female participation rate (19.64%).

Overall it could be said that in the entire valley, women's participation is quite low in spite of a growing literacy rate. The efforts of the National Literacy Mission and Sarbho Shiksha Abhijan in propagating awareness of literacy may also be taken as a positive factor in promoting literary awareness. Male literacy, which is only marginally higher (65.74%) than the female literacy (63.92%), reflects a comparatively higher participation of male workers (57.27%) in Barak Valley, and this general trend is observed for all the districts. This shows that gender discrimination exists, not only in terms of literacy but also with regard to participation rate among males and females. On the whole the findings that emerge from our study is that, higher literacy rate has raised the status of women in the valley but this has not made entry in the labour market any easier for women.

Decision-making

Any form of national and social policy which has an impact on women's work, income and welfare cannot be implemented without at the same time enlarging women's participation in decision making [Report, 1974-1978]. Apart from social traditional, cultural norms that dictate the individuality of a woman there are several semi-legal and quasi-legal practices that tend to erode the decision making power and the legal position of the women within the family. Such submissive nature predominant among women flows from concepts such as male guardianship. They affect women's autonomy and authority. Therefore, the ability to take decisions as regards family affairs is vital for the smooth functioning of not only a particular household but also the society as a whole and may therefore have a favourable impact on their WPR.

Women who have greater autonomy of taking decisions may also decide whether they should go outside the confines of their household activities and work along with their male counterparts. They also have greater autonomy in deciding the nature and type of work that they will undertake. Since the power and ability to take decision is associated with enhanced status of women, it also implies that such women may have a say when they are discriminated in their workplace.

In order to judge the capability of women in decision-making process we have taken mainly three heads viz., the decision to spend money, decision taken in the purchase of household goods and decisions regarding family affairs and children. Responses to these queries help us to asses the financial independence, power and authority assigned to women within the household. Though there are other wide range of variables influencing a woman's decision making ability, but due to limitations of time and sample size, we have restricted to three variables for the purpose of our study. The details of decision-making as recorded during our investigation are in Table 1.5.In Barak Valley of Assam, a glimpse into Table 1.5 indicates that in all the urban areas of the district of Cachar, Karimganj and Hailakandi women enjoy a greater autonomy in decision-making capability than women in the rural areas, though greater freedom is envisaged in matters of household consumption than spending money.

Table 1.5

Decision-making Level of Female (in %) in different household activities

Sl. No.	*Valley/ District*	*On the basis*	*Spending money*	*Purchase of house hold items*	*Family & children affairs*
01	Barak Valley	Total	62.6	88.9	61.1
		Urban	84.1	93.2	88.6
		Rural	58.7	89.0	55.5

(Contd...)

Sl. No.	*Valley/ District*	*On the basis*	*Spending money*	*Purchase of house hold items*	*Family & children affairs*
02	Cachar	Total	60.3	88.7	54.6
		Urban	72.7	100	90.9
		Rural	60.0	89.2	51.5
03	Karimganj	Total	63.3	88.6	69.6
		Urban	84.6	84.6	84.6
		Rural	59.1	89.4	66.7
04	Hailakandi	Total	69.0	90.5	66.7
		Urban	90.0	95.0	90.0
		Rural	50.0	86.4	45.5

Source: derived from Paul. R. "Structure of women's participation rate and its determinants in Assam with special reference to Barak Valley." Ph.D. thesis (unpublished), Assam University, Silchar, 2002.

In urban areas of Cachar women enjoy almost cent percent liberty in matters pertaining to household consumption. In both the districts of Cachar and Hailakandi almost 90% of the decisions regarding education of the child and family related decisions are taken up by the women. But the women interviewed have expressed men folk are also consulted. Urban FWPR is also high in all the districts of Barak Valley except Karimganj where FWPR is high in the rural area according to the field survey. Increased literacy in urban areas added with it the greater demand of urban life automatically entrust women to share greater responsibility and this also entails greater decision-making process in the urban areas.

Rural FWPR is high in all the districts except Hailakandi where FWPR is high in urban area. Increased FWPR in Hailakandi is also followed by increased decision making in all the categories of decision-making. In Cachar too, women enjoy sufficient independence with respect to decision-making capabilities. In spite of this, FWPR is low in all the districts of Barak Valley as compared to male counterpart and the participation situation is not at all encouraging. It could be thus said that while many women are capable to take decision within the household, such decision exercising power do not always gainfully translate into jobs in the formal sector. The table further reveals that though on an average almost 60% women in Barak Valley are able to decide as regards spending and the interest of their children but the rest 40% do not posses such autonomy. Absence of taking independent decision follows from patriarchal ideology and subordination of women. Interviews with the rural and urban women have revealed that at the community level, very few women have access to decision making, in all the three districts surveyed,

Attitude of males towards women's work as a factor determining FWPR

Attitudes towards work are systems of meaning that are contextually derived throw light on thosc contents and lead to direct them [Denzil, 1990]. Participation in the work force is considered to be an important factor in determining the position of an individual in the family as well as in society. It determines the level of food availability, nutrition and level of other essential demands. It holds the key to productivity [Mitra, 1973]. A positive attitude to women's work entrusts some degree of autonomy to a women's status. Among lower-socio-economic groups it allows for greater mobility which enables them to compete for greater economic independence. Sociologists point out that control over income and freedom to work does not always confirm power and independent status since patriarchal ideologies govern the every construction of women.

Interview with the male members of the sample households revealed several factors as regards the refusal or inhibition of women working outside their domestic activity. In our survey a very satisfactory trend has been observed with respect to male's attitude towards women's work. In the overall analysis urban men depict greater positive attitude towards women's work than in the rural areas (90.5% in urban areas as compared to 87.5% in the rural areas) (Table 1.6). Among the three districts, Karimganj depicts a higher percentage of favourable attitudes of men towards women's work both in the urban as well as in the rural areas perhaps due to greater literacy in the district next only to Cachar. The rural areas of Hailakandi show a trend more satisfactory than the urban areas of other districts. Improved sex ratio in the valley coupled with greater literacy in the Valley is indicative of an improved outlook of men. With unemployment - real and disguised – acute in the valley, participation of women in outside work is perhaps seen as a necessity for the upliftment of the economic position of the households.

Table 1.6

Attitude of Male towards Female Work (in %)

Sl. No.	*Valley / District*	*Positive Attitude of Male towards Female Work*		
		Total	*Urban*	*Rural*
01	Barak Valley	88.2	90.5	87.5
02	Cachar	85.7	90.9	85.1
03	Karimganj	91.1	92.9	90.5
04	Hailakandi	90.6	88.2	93.3

Source: derived from Paul. R. "Structure of women's participation rate and its determinants in Assam with special reference to Barak Valley." Ph.D. thesis (unpublished), Assam University, Silchar, 2002.

Average household size, monthly income level and monthly expenditure level

Household size represents the number of members in the family. The larger the number of persons in the household, the greater the household's taste (or "need") for money income. and therefore the greater urgency can be expected for members of the family to be absorbed into participation process. The greater family responsibilities suggest that women may have a stronger propensity for market work. Relatively larger size of the household is due to varied factors. Like more average children per household resulting in higher responsibility for women, existence of a joint family set-up, which restricts the participation of women, more of earning male members also acts as a deterring factor for women to go into the public sphere and work. All these factors create to an obstacle for women to participate. Our field finding has been tabulated in Table 1.7

In our present analysis, as shown in the Table 1.7, the household size is found to be lower in urban areas varying between 5.3 and 5.8 in all districts, whereas for rural areas it is marginally higher varying between 5.7 and 6.2. The average size of the household needs to be examined in relation to average monthly income and expenditure pattern of the household. With high per capita income large household size is compensated and women may not show much inclination or may not be forced by economic necessity to work in urban as well as rural area. On the other hand it is also seen that large family size compels women to supplement the family income particularly in rural area. In our analysis in all the districts of Barak Valley, average urban income as well as expenditure is found to be higher than average rural income and expenditure. Thus it may be considered that mere household size does not contribute in the determination of female participation in work.

Table 1.7
Average household size and income & expenditure

Sl. No.	*Valley / District*	*On the basis of*	*Avrg House hold size*	*Average Monthly Income*	*Av. Monthly Expenditure*
01	Barak Valley	Total	5.80	Rs. 6077.10	Rs. 4251.54
		Urban	5.49	Rs. 11229.00	Rs. 6691.98
		Rural	5.89	Rs. 4966.93	Rs. 3604.53
02	Cachar District	Total	5.90	Rs. 5965.88	Rs. 4070.15
		Urban	5.80	Rs. 13977.50	Rs. 8355.00
		Rural	5.90	Rs. 4890.77	Rs. 3444.81
03	Karimganj District	Total	5.60	Rs. 6324.00	Rs. 4517.14
		Urban	5.40	Rs. 10591.43	Rs. 6971.43
		Rural	5.70	Rs. 5257.14	Rs. 3903.57
04	Hailakandi District	Total	5.80	Rs. 5971.54	Rs. 4342.22
		Urban	5.30	Rs. 8494.00	Rs. 5385.78
		Rural	6.20	Rs. 4395.00	Rs. 3690.00

Source: Derived from Paul, R. "Structure of women's participation rate and its determinants in Assam with special reference to Barak Valley," Ph.D. thesis (unpublished), Assam University, Silchar, 2002.

Income and Participation of Women

The lower the socio-economic status of the family, the greater the proportion of total income contributed by women. Majority of the women workers in the sample under study come from very poor families with the average income per month is as low as rupees 500 per month. This amount though small is essential for the survival of the family. Therefore, it could be argued that the decision to work would to a large extent depend on the relative levels of the family income. Given the fact that only 22.89% are engaged in various productive process in the entire Barak Valley, it is observed that almost 50% of the female workers earn a monthly income varying between Rs. 500 and Rs.2000. This is considerably low. Only 2.7%, 2.2% and 0.5% earn an income varying between Rs. 1001 to Rs. 15000. Approximately on average women contribute almost 25% to their total family income. But most of these works is dismissed as secondary work and in majority of the cases most of this amount is used for consumption of non-essential commodities.

The monthly per capita consumer expenditure reflects the average economic level of the state and is also a very important determinant for a women's decision to work. As economic level of the household improves, there is a greater tendency for women to gradually withdraw from the labour force, because men prefer women to tend the children and look after the household affairs. In the entire Barak Valley, the average monthly expenditure is high in all the urban areas spread over the three districts. Female work participation is however comparatively low in all the urban areas except the district of Hailakandi. In this connection it could be mentioned that in the urban areas, participation is mainly concentrated in the organized sector, which is better paid with higher level of income. Moreover, in urban areas there is greater provision of people being engaged in secondary occupations enabling them to double their income. Urban life necessitates better standard of living. With higher level of income, better standard of living and provision of additional income; all these factors jointly contribute to an increase in family expenditure of a household. In the rural area, though family expenditure is low income is also low. With a high participation of females as well as males, female participation even if remunerated is much lower than men.

The Way Ahead

To achieve gender equity, it is imperative that gender dimensions of development be looked into. A sensitive and a comprehensive dynamic policy needs to be framed to include women as active partners in the development process.

Conclusion

Overall it could be said that female workers in Barak Valley like the rest of India are restricted to marginal status in the workforce and gender participation in home-based and traditional production needs to be comprehensively estimated and accounted which could ensure a better picture of women's participation rate in the valley. This regional pattern of low female work participation rate as examined in Barak Valley is consistent with the character of gender relations in different parts of the country. The north-western states are distinguished for their highly unregulated gender relations and distribution of work patterns between males and females. High literacy rate, which is a characteristic feature of the state of Kerala, also has a distinguishing history of a liberated society. Kerala's particular experience of high literacy and improved sex- ratio is quite remarkable and has been an extremely important lesson for the rest of India. This has in return enhanced the social standing of women in the state and has brought about significant changes in terms of social achievements, less unequal gender relations, less patriarchal kinship systems, less male dominated property rights and greater prominence of women in influential social, political and economic activities. Not far away, the economic role played by the women of Manipur has influenced the social progress of the women of Manipur. Similarly the rapid transformation of the women of South Korea and China brought about by enhancing the educational status of women and equipping them with necessary training and skill has created a transformation in the participation of women in the economic sphere. The success of these states and countries could be taken as an example to illuminate the social standing and position of the women of Barak Valley, which also dominates in aspects of literacy among females and high sex ratio. These regional contrasts provide a useful means of investigation in different aspects of women's participation in the valley and these experiences are but important lessons to emulate and adopt them in practice.

REFERENCES

Agarwal, Bina (1995), *A Field of One's Own Gender and Land Rights in South Asia.* Cambridge University Press.

Banerjee, N. (1995), *Women's Employment and All That, The Administrator,* Vol XL, Jul-Sept 1995.

Banerjee, N. (1999), 'How Real is the Boogey of Feminization', in Papola and Sharma (eds), *Gender and Employment in India*, Vikas Publishing House, New Delhi.

Basu, Sibtapan (2000), *Barak Upatyakar Mati O Manush*, Sungraphics, Silchar.

Beneria, Lourdes (1978), *Production, Reproduction and the Sexual Division of Labour,* ILO.

Biswas, Anurupa (1998) : *Barak Upttakar Nari Jagoron,* Pragati Publishers, Silchar.

Boserup, Ester (1970), *Women's Role in Economic Development*, George Allen and Unwin, London.

Chatterjee, Suhas (2000), *A Socioeconomic History of South Assam,* Printwell Publishers Distributors, Jaipur.

Denzil, Saldanha (1990). 'The Socio-Economic Context of the Warli AttitudeTowards Work' in A D Moddie (ed), *The Concept of Work in Indian Society*, Manohar Publications, New Delhi.

Dutta Ray, B. (1978), *Social and Economic Profile of N.E. India,* (ed), B.R. Publishing, Delhi.

Economic Intelligence Service (1998), *National Income Statistics*, CMIE.

Fong, M.S. (1982), *Measuring Women's Work in Agriculture,* Seminar Paper, Institute of Social Studies, New Delhi.

John E. Mary, (1996), 'Gender and Development in India, 1970's – 1990's some reflections on the constitutive role of content's,' *EPW*, Vol. xxxi.

Mitra. J. (1997), *Women and Society, Equality and Empowerment*, Kanishka Publishers, New Delhi.

Mitra, A. (1973), *The Status of Women : Literacy and Employment*, Allied Publishers, New Delhi.

Singh, Okram Rudra Babu (1994), *Silk Industry in Manipur*, NEIHA, Shillong.

Paukerl, Liba (1984), *The Employment and Unemployment of Women in OECD Countries*, OECD.

Report (1974-1980), *On Women Workers and the New International Economic Order*.

Sarma Satyendra Nath (1989), *A Socio-Economic and Cultural History of Medieval Assam (1200-1800 A.D)*, Guwahati.

Sinha, R.C. (1979), "Agricultural Development and Rural Employment," in Papola, T.S. etal.(ed), *Studies on Development of U.P.*, Giri Institute of Development Studies, Lucknow.

Statistical Profile of Barak Valley, NECAS (1998), *An Agenda for Research on the Valley's Economy*, Silchar.

Sudarshan, Ratna M. (2001), 'Empowerment of Women: The Economic Role', *Yojana*, vol. 45, Aug, 2001.

Usha, S. (1981), "Male-Female Wage Differential in an Urban Rural: A Case Study of Madras Metropolitan Area", Unpublished Doctoral Thesis, Madras University.

Vanamala, M. (2001), 'Informalisation and Feminization of a Formal Sector Industry': A Case Study', *EPW*, Vol xxxvi, June 30.

9

Social Sector Discourse in Development Perspective of Barak Valley

—Hiranmoy Roy

The term 'social sector' refers to education, health and nutrition. The term has not been formally defined and thus has come to several connotations. There are two main approaches to the definition which may be identified from human resource development view and the human development approach respectively. The first approach focuses on investment in education, health and nutrition as a means to augment the quality of human capital which is defined as the 'stock of skills and productive knowledge embodied in people.' This approach drew attention when Becker [1998] highlighted that education, health and nutrition cannot be considered merely as consumption goods. Investment in these sectors is emphasized as a means to higher productivity of the labour force and is justified on the basis of either financial rates of return as in the case of education or estimates emerging from production functions as in the case of health. The human resource development approach does not take poverty as a factor that limits the ability of individuals to invest in them. On the other hand, the human development approach has been defined by UNDP as 'the process of enlarging people's choices' [UNDP, 1990]. The concept encompasses empowerment, co-operation, equity in basic capabilities and opportunities, sustainability and security. In this approach people occupy the centre stage and the measures such as education, health and nutrition are emphasized for their intrinsic value and for their role in enhancing basic capabilities of people.

The emergence of the concept of human development can be traced to the efforts of Professor Amartya Sen who defined standard of living in terms of functioning and capabilities rather than in terms of commodity possession. This approach treats human beings as ends in themselves, whereas the human resource development approach treats people as human capital that leads to greater human prosperity. In other words, human development approach perceives human beings as ends in themselves as opposed to the human resource approach which emphasizes the means or productivity aspect of human beings. In the human resource development approach investment in human capital is governed by the rates of return on such investment. In the human development approach, the acquisition of education, health and nutrition is considered as an essential human right and is promoted even if the conventionally measured rates of return on these investments are low. The human resource development approach places considerable emphasis on individual initiative to further the cause of education, health and nutrition. In sharp contrast, human development approach advocates a higher importance to the supply of social services, particularly public services. The use of this concept also implies a greater role and responsibility for public provision.

Despite the distinctions, the two concepts are linked together by the fact that human development provides the preconditions for human resource development which in turn contributes to economic growth. There are number of studies [Dreze and Sen, 1989] emphasizing upon the role of economic growth and state intervention in the process of human development. Economic growth helps in human development through reduction in poverty, improving access to basic social services and building up of basic capabilities in the people [Taylor, Mehrotra and Delomonica, 1997]. The process of human capital formation contributes to human development and thereby to economic growth as human development not only enhances well being or quality of life of the people, but it also adds to productive capacities of people. Therefore, economic growth and human development are interdependent. The activities covered under social sector contribute to human capital formation and thereby to human development and are recognized as critical inputs in improving the index. Social Infrastructure is as critical for human resource development as for physical infrastructure [IGIDR, 2002].

Social sectors may be defined as those providing social security or the protection which the society provides for its members through a series of measures, like the provision of medical care and provision of subsidies for families with children, against the social and economic distresses that otherwise would be caused by substantial reduction in earnings resulting from sickness, maternity, employment injury, unemployment, invalidity, old age and death. Social security essentially is an objective to be pursued through public means. Different heads of social securities are :

1. Promotional Social Security

    (i) Education, Art, Culture

    (ii) Medical, Public Health, Water Supply and Sanitation

    (iii) Family Welfare

(iv) Housing

(v) Urban Development

(vi) Labour and Employment

(vii) Social Security and Welfare

(viii) Food Subsidy

(ix) Rural Development

2. Protective Social Security

(i) Relief on account of Natural Calamity

Social Sector = (1+2)

The protective social security is concerned with preventing a decline in living standards in general and in the basic conditions of living in particular. The problem of protection is paramount in the context of famine prevention, and also in dealing with other kinds of sudden economic crises and sharp recessions. On the other hand, the objective of promotional social security is to enhance the normal living conditions and dealing with regular and often persistent deprivation. It is in a sense more ambitious in wanting to eradicate problems that have survived thousands of years. The strategic issues involved in promotional social security may differ very considerably from those of protective social security. The policy requirement for augmenting social sector performance are (i) Policies relating to land (ownership and tenancy reform), (ii) Policies relating to employment, and (iii} Direct public provision of basic needs such as food, housing, health care, etc.

To study the performance of social sector, it is imperative to examine the trends of social sector outlays of the government and to assess their impact on social sector.

Table – 1

Revenue and Capital Expenditure of the Central Government during Pre Reform (1981-82 to 1991-92) and Post Reform periods (1991-92 to 2000-01) (as % of expenditure)

Period	*Revenue Expenditure*			*Capital Expenditure*			*Aggregate expenditure as % of GDP*
	Plan	*NP*	*Total*	*Plan*	*NP*	*Total*	
Pre-Reform	13.16	53.96	67.12	20.98	12.41	33.39	21.10
Post-Reform	14.02	63.78	77.80	10.80	9.87	20.67	20.36

Source : (i) CMIE, Public Finance, Economic Intelligence Service, March 2002.
(ii) Govt. of India, Economic Survey, Various Issue.

From Table-1, it is evident that aggregate social sector expenditure as a percentage of GDP has fallen in post-reform period to 20.36% as compared to pre-reform period of 21.10%. Both plan and non-plan expenditure in revenue category as a percentage of GDP has increased, while there is sharp fall in plan and non-plan expenditure in capital expenditure category. Thus fall in aggregate social expenditure is primarily due to fall in capital expenditure. To get a clearer picture of the social sector performance segregated data on central and state government developmental expenditures are shown in Table-2.

Table-2

Aggregate Development Expenditure of Central and State Governments as percentage of their Average Total Expenditure

Period	*Central Govt.*	*State Govts.*	*Combined*
Pre-Reform period (1980-91)	54.7	70.2	61.8
Post-Reform period (1991-2000)	45.8	63.2	53.3

Source: Computed from figures in *RBI, Hand Book of Statistics on Indian Economy*, 2000, p. 142.

It can be observed from Table-2 that there is sharp reduction in central as well as state developmental expenditures during post-reform period compared to that of pre-reform period. It is very important to observe that the state governments' developmental expenditures have fallen from 70.2% to 63.2% in between pre-reform and post-reform period, which has an adverse affect in the state of Assam as well.

To analyse the performance of social sector in the State of Assam, some selected indicators viz. per capita NSDP, HDI, growth of GSDP, infant mortality, crude death rate, life expectancy, literacy rate, per capital income, etc. are taken as representative of social sector performance. Growth of Gross State Domestic Product declined to 2.7% in the year 2000-01 compared to 3.6% in 1990-91. To compare the relative position of Assam compared to that of Kerala some specific indicators, viz. infant mortality rate, life expectancy and literacy rates are adopted in this study. The relative position of Assam is not satisfactory. The state of Kerala is ahead of Assam as shown by the indicators of quality of life in Table-4.

Table – 3

	2000-2001	*1991-92*
Per capita NSDP	Rs. 9170	5685
Per capita Rank	10	12
HDI	0.367	0.348
Rank	15	10
Growth of GSDP	2.7	3.6
Growth of per capital GSDP	1.0	1.4

Source : Govt. of India (2002), Tenth Five Year Plan.

Table – 4

Indicator	*India*	*Kerala*	*Assam*
Infant Mortality Rate (per 1000)	106	37	116*
Crude Death Rate (per 100)	12	7	
Life Expectancy	55	66	
Literacy (%)	29	70	
Per capita income (US$)	150	225	

Source : Govt. of India, 1985, Vol.-2.

Table – 5

Literacy	*1951*	*1961*	*1971*	*1981*	*1991*	*2001*
Assam	18.53	32.95	33.94	-	52.89	63.25
India	18.33	28.30	34.45	43.57	52.21	64.84
Kerala	47.18	55.08	69.75	78.85	89.81	90.86

Source : Economic Survey, 2006-07.

Table – 6

Life	*1961*			*2003*			*2005*		
Expectancy	*M*	*F*	*Person*	*M*	*F*	*P*	*M*	*F*	*P*
Assam	NA	NA	NA	69	65	67	66	69	68
India	122	108	115	57	64	60	56	61	58
Kerala	55	48	52	11	12	11	14	15	14

Source : Economic Survey, 2006-07.

The striking differences between the states of Assam and Kerala are that Kerala experienced more effective land reforms and in respect of public provisioning of basic needs - food, health care and education - Kerala stands out with much higher rates of success. The studies relating to social sectors performance in the Barak Valley is more disappointing. Cachar, Karimganj and Hailakandi - the three districts in the Barak Valley of Assam - exhibit one of the lowest levels of economic development in the North-Eastern Region. This calls for greater emphasis on the issues of economic development of these districts and placing much more attention on the social infrastructure, more specifically social sector issues. There is clearly lack of efforts to enhance the levels of human development in these districts.

To evaluate the social sector performance in Barak Valley, an analysis of the social sector indicators are necessary. This presentation makes an attempt to analyse the social sector issues in Barak Valley. For the purpose of analysis, the two districts, viz. Cachar and Karimganj, were selected. The relative positions of the districts based on different social sector indicators are presented in Tables-7, 8 and 9.

Table – 7

	Cachar	*Karimganj*
HDI	0.402	0.301
HDI Rank	08	19
GDI	0.409	0.012
GDI Rank	14	23

Source: Assam Human Development 2003.

Table – 8

	Cachar		*Karimganj*	
	1991	*2001*	*1991*	*2001*
Child Mortality Rate (upto 5 yrs.)	116		131	
Crude Birth Rate	32.61		48.52	
No. of Hospitals		8		2
PHC's		22		16
Dispenseries		2		5
No. of Beds (per thousand population)		6.93		2.5

Source: Assam Human Development Report 2003.

Table – 9

	Cachar		*Karimganj*	
	1991	*2001*	*1991*	*2001*
Literacy rate (%)	59.16	68.42	54.71	67.21
Male literacy rate	68.8	76.51	64	73.9
Female literacy rate	58.8	59.85	44.8	60.09
Scheduled Caste				
Male literacy rate	64.5		60.49	
Female literacy rate	46.7		45.86	
Scheduled Tribe				
Male literacy rate	64.5		35.48	
Female literacy rate	46.7		20.33	

Source : Assam Human Development Report 2003

Table – 10
Literacy by Income Groups

	Persons	*Male*	*Female*
Cachar	58.0	65.8	55.3
Up to 20,000 (low)	51.2	53.2	48.2
20,000 – 40,000 (Lower middle)	53.4	54.8	49.5
40,000 – 62,000 (Middle)	57.2	58.2	53.2
62,000 – 86,000 (Upper middle)	59.4	59.4	55.1
86,000 and above (High)	62.3	68.3	57.2
Karimganj	56.8	62.3	51.0
Up to 20,000 (low)	48.2	51.2	46.2
20,000 – 40,000 (Lower middle)	51.4	53.4	48.4
40,000 – 62,000 (Middle)	53.2	55.8	49.8
62,000 – 86,000 (Upper middle)	55.3	57.5	53.2
86,000 and above (High)	58.6	59.3	54.5

Source: Survey conducted by the author.

Table – 11
Major Morbidity (per thousand populations)

	Diabetes	*Heart diseases*	*Cancer*
Cachar	25	42	15
Karimganj	28	40	18

Source : Survey conducted by the author.

Table – 12
Percentage of Children Immunized

States / Districts	*Percentage of children immunized*
Cachar	60.2
Karimganj	58.1

Source: Survey conducted by the author.

Table – 13
Availability of Drinking Water Facilities by Districts

	Piped water	*Other protected*	*Unprotected*
Cachar	20.3	52.8	26.9
Karimganj	18.1	43.7	38.2

Source: Survey conducted by the author.

Some additional information on literacy, health and drinking water facilities in these districts based on primary data are shown in Tables – 10, 11, 12 and 13. The primary data were collected following the multi-stage stratified random sampling.

Both the districts of Cachar and Karimganj are placed at lower human development index as shown in Table-7. There is huge difference between the child mortality rates in Kerala and in these two districts of Barak Valley in Assam. Child mortality rate is a very important indicator of health, but the relative position of the districts under study shows that Barak Valley districts are far behind the advanced states like Kerala.

In terms of literacy rate these two districts of Barak Valley are in a better position in Assam, but still lagging far behind Kerala. From Table-10, it can be observed that literacy rate in the Barak Valley districts are higher among the high income groups and male literacy is higher than that of female literacy. The health services in the districts are quite unsatisfactory as revealed from Table-8. The number of hospital beds per thousand of population in two districts of Cachar and Karimganj are 6.93 and 2.58 respectively. The percentages of people not having safe drinking water in Cachar and Karimganj districts are 26.9 and 38.2 as shown in Table-13 and it represents a very low level of quality of life. The position of Hailakandi district in terms of social sector performance stands in between Cachar and Karimganj and exhibits no sign of radical difference from other two districts.

Conclusion

The social sector deprivation in Barak Valley is indeed acute. The task of providing minimum level of social security is, therefore, not going to be easy one. This will require sound strategy; suitable policies and programmes and specific targeting of different sectors of the society. There is an imperative need to strengthen political, financial and institutional commitment to enhance social sector aspirations in this valley. The land reform should be an urgent agenda, although it will be very difficult to implement policies relating to control over land due to complex nature of landholdings in all three districts of Barak Valley. In the employment sector, it will be necessary to create appropriate environment for self employment within the valley, since both the public and the private sectors are awesomely week and non-responsive to the needs and aspirations of the masses. More importantly, stress has to be laid on public provisioning of basic needs such as food, healthcare and education. Finally, higher allocations and proper delivery mechanism to the beneficiaries can go a long way in enhancing social sector achievement in the Barak Valley.

REFERENCES

Bhanushali, S.G. (2007), *Poverty Scenario in Indian States*, Bookwell, New Delhi.

Backer, L Therese (1998), *Doing Social Research*, McGraw Hill Inc., New Delhi.

Dev, S. Mahendra and Mooij Jos (2002), "Social Sector Expenditures in the 1990's : Analysis Central and State Budgets", *Economic and Political Weekly*, 37(9).

Dev, S. Mahandra and Mooji Jos (2003), *The Politics of Social Sector Expenditure: An Exploratory Analysis of the Budget Making Process*, IDS Working Press, Sussex.

Dreze, J. and Amartya Sen (1989), *Hunger and Public Action*, New York: Oxford University Press.

Government of India (1946), *Report of the Committee on Health Survey and Development*, New Delhi.

Government of India (1986), *National Policy on Education*, New Delhi.

Government of India (1989), *National Health Policy*, New Delhi.

IGIDR (2002), India *Development Report, 1999-2000*, Oxford University Press.

Joshi, Seema (2003), "Social Sector: Need for Reprioritization", *Southern Economist*, September 15.

Kumar, G. (1987), *Well-being, Deprivation and Basic Needs : A Perspective on the Development Experience in Kerala*, Oxford University Press.

NCAER (2002), "Development of a Macro Economic Model to Assess the Impact of Social Sector Spending in India", Project Sponsored by Canadian International Development Agency.

Pant, S K (ed.) (2006), *Human Development, Concept and Issues in the Context of Globalization*, Rawat Publication, Jaipur.

Parr, Sakiko Fukuda and A.K. Shiva Kumar ed. (2007), *Readings in Human Development*, Oxford University Press.

Prabhu, S. (1994), "Social Expenditures in India: Trends and Implications", Background paper on UNDP.

Prabhu, S. (2001), *Economic Reform and Social Sector Development*, Sage Publication, New Delhi.

Sarkar, Siddhartha (2007), *Gender, Work and Poverty*, Serials Publication, New Delhi.

Sen, A.K. (1981), "Public Action and the Quality of Life in Developing Countries", *Oxford Bulletin of Economics and Statistics*, 13.

Taylor, Mehrotra and Delamonica (1997), "The Links Between Economic Growth, Poverty Reduction and Social Development: Theory and Policy" in Mehrotra and Jolly (eds), *Development with a Human Face : Experiences in Social Achievement and Economic Growth*, Oxford University Press.

10

Border Trade and Growth Rate in Economy of Barak Valley

—Ramanuj Chakravorty

The concept of border is still heavily tinged with the notion of exclusive territorial sovereignty and, therefore, the term 'border' is a confusing and complex term that is difficult to define. The scholars from different disciplines have tried to define the term according to their own perception, everyday experience, heritage and cultural background or occupational biases. In fact, border refers to a place where human groups live on both sides of the dividing line between two countries up to a given point. One important point that deserves reference is the heterogeneity of the borders of the different countries. Thus, in each border the situation is different from another. Each one has its own peculiar problems and varying degrees of coordination and integration - internally with its own country and externally with the neighbouring country.

Border Trade

The term 'Border Trade' has different connotations in different contexts. Although the border trade is nothing but international trade, it has some distinct features. It takes place between the peoples in the bordering areas of the two sides, and the participation of the people and institutions from mainland of the countries remain almost absent. Thus, border trade is the carrying out of trade in the bordering areas with full consideration of the existing economic and social situations without disturbing the delicate socio-economic balance that has emerged as a result of changes in the people's life style and in the orientation of the economic activities of two different national systems. The changes that have taken place over the ages because of divergence in the monetary system, tax, labour, migration etc and the policies that each nation practices do significantly influence the character of the border trade Nonetheless, this definition of border trade is only tentative. Border trade today means much more than what has been perceived above. The border trade or cross-border economic activities within different regions have changed significantly over the last decades and these are becoming a more effective and comprehensive form of technological and cultural cooperation. Thus, change in the pattern of border trade has occurred throughout the world. The European Union (EU) today has emerged as a complete borderless economic union. In early 1990s the governments of Vietnam and its neighbouring countries made great effort to restore bilateral economic relation which was almost destroyed by the border war in 1979. A number of 'land border gates' resumed their functions and operations leading to increased trade and other economic activities between peoples living along the borders. Gradually these activities have officially been recognized by the respective countries as an important part of their national economy.

In the global context it may be observed that irrespective of the forms and the levels of border trade that the countries are practicing or experiencing, there are certain common features in the economy of the bordering areas of almost all the countries:

1) Bordering regions exhibit low level of economic development than the mainland of the countries.

2) Border regions are generally richly endowed with natural resources which remain untapped and unexploited, perhaps because of the indifferent attitude of the policy makers from the mainland of the countries.

3) Subsistence economic conditions have been a long tradition of border residents.

4) Border regions are inhibited by ethnic minorities.

5) It is not unknown that due to the existence of centrally oriented development models, which have always considered the national borders as 'reserve areas' that can be incorporated at any time in future, the bordering areas in most of the countries are peripheral and marginal spaces that enjoy little or no integration with their respective national economies and societies. Economic, social and even cultural problems in the bordering areas are much more intense than many rural areas in the interiors of many countries.

6) Policy makers keep on postponing the formulation of plans that would contribute to the economic take off of the bordering regions which calls for total involvement and active participation of the local people.

Border Trade and Barak Valley

Barak Valley (or Southern Assam) because of its peculiar geographical location has been in the periphery and far off from the mainland of the country and the heartland of the state of Assam. It borders Bangladesh, and the tiny valley is a 'border region' in every sense of the term. Like many other bordering areas, the Barak Valley region of Assam is economically underdeveloped. In fact, economically it is more backward than many other bordering areas of the country. However, today there is enormous scope for every bordering area, including the Barak Valley, to exploit the geographical closeness and proximity with the neighbouring country through border trade. The geographical isolation of the valley, which was once considered as one of the reasons for underdevelopment, can now become an opportunity for generating growth impulses in local economy through border trade with Bangladesh. Therefore, border trade should be given emphasis for registering growth impulses in the economy of Barak Valley which is now not very encouraging.

The border trade or cross-border economic activity between Bangladesh and the Barak Valley has started drawing attention of the policy makers only after 1990s. Before that the Barak Valley and the whole of Northeast figured in the policies of the Government of India only in the context of national security. The role of border trade in the economy of the region could not then attract the attention of the national policy makers. Therefore, the region was not able to find a place in the industrial map of the country. The hostile political relations with the neighbouring countries also came on the way of cross-border trade. The things have changed now. The Northeast now figures prominently in India's policy towards Southeast Asia and the country's initiative towards integration of national economy with the world economy (popularly known as globalization of Indian economy). Bangladesh also occupies an important place in India's international trade strategy since India considers Bangladesh as one of the important destinations of country's exports. A substantial part of the Indo-Bangladesh trade has to pass through Barak Valley because this valley shares a common border with Bangladesh, the socioeconomic ties between Bangladesh and this valley is age old, and some of the old communication links can be easily activated in spite of the fact that the partition of India had disrupted those links.

Border Trade through Karimganj

Table 1

Export and Import Trade between Land Custom Stations of Barak Valley and Tripura, and Bangladesh

Name of the Land Custom Station	*Year*	*Value of Export (in rupees)*	*Value of Imports (in rupees)*
Agartala (Tripura)	2004-05	nil	110769327
Agartala	2005-06	4877179	276058712
Agartala	2006-07	2604838	467043182
Agartala	2007(April-October)	6281224	333386169
Karimganj Steamerghat Ferry Station (KSFS)	2003-04	19933415	47387
KSFS	2004-05	36007575	933229
KSFS	2005-06	26450304	1163038
KSFS (up to 31.12.06)	2006-07	22836990	568675
Sutarkandi	2004-05	213372113	—
Sutarkandi	2005-06	322733572	100000000
Sutarkandi (up to 31.12 06)	2006-07	326297696	270000000

The above table reveals some information about the export and import trade with Bangladesh through Agartala (Tripura) and Karimganj (Barak Valley, Assam). The table clearly shows that imports from Bangladesh through Agartala have been increasing very fast. From Rs 110769327 in 2004-05 it went up to Rs 467043182 in 2006-07 and it might touch 50 crores in 2007-08 because the value of imports for the seven months period (April-October) was more than Rs 33 crores. Thus, the rate of increase in imports has been more than 200% since 2004-05. However, the rate of increase in exports through Agartala is lower than Karimganj. Exports through Agartala actually came down in 2005-06 from what it was in the previous year, although in 2007 it showed an upward trend. On the other hand, the rate of increase in exports from both the land custom stations of Karimganj has been more than 50%. If we go by the growth trend for eight months period, it seems that exports would remain stagnant in 2006-07 when this year's volume of exports is compared with that of the previous year. This might be because of the ongoing political uncertainty in Bangladesh. The rate of growth in imports from Bangladesh through land custom stations of Karimganj is also showing upward trend. But there is a variation in the volume of trade between the two stations in Karimganj. The volume of both imports and exports from Sutarkandi Land Custom Station of Karimganj is much higher than what it is in case of Karimganj Steamerghat Ferry Custom Station. However, one notable feature of the border trade between Karimganj and Bangladesh is that the two Barak Valley stations have been experiencing surplus in the balance of trade with Bangladesh. Another point which deserves attention is that the volume of illegal trade from this part of the country with Bangladesh has become insignificant in the wake of globalization.

The commodity structure of exports from Barak Valley consists of coal, marble tiles, stainless steel, distillation plants, sanitary items, motor parts, limestone, dry chilly, ceramic wares, fresh ginger, fresh fruits, orange, citrus, betel leaves, fresh vegetables, non-basmati rice, chaw-chaw, varieties of dry fish, etc. Imports from Bangladesh consist of molasses, deep fried crispy snacks, cheese ball, potato cracker, ring chips, lachha, chanachur, etc. If the commodity structure of export is analyzed it can be observed that it is composed of the products from both primary and secondary sectors. It means that if proper plans for industrialization of the Barak Valley are drawn with due emphasis on agriculture then the valley does not need to depend on marketing of its raw-materials only. The products which are having demand in Bangladesh markets are cement, automobiles, hardware, marble tiles, etc. These products come from West Bengal, Maharashtra, and Gujarat and find their way to Bangladesh through Karimganj. The initiative on the part of the Government of India to develop manufacturing industries for these items in Barak Valley and the neighbouring areas is sure to be cost effective and economically rewarding since there is a ready market in Bangladesh. The paper is manufactured in the Barak Valley through public sector unit of Hindustan Paper Corporation at Panchgram and is also being exported to Bangladesh. The export of paper is managed by some public sector export houses. The coal exported to Bangladesh through Karimganj is sizeable, but this is actually extracted from the coal mines of Meghalaya. But the economy of Barak Valley cannot gain much by facilitating the export only. Therefore, a product line composed of both agricultural and industrial sectors should be developed on the basis of demand in Bangladesh markets. The demand for non-basmati rice has been on the rise in Bangladesh, but after 1998-99 there was no export of non-basmati rice through Karimganj border. The message is loud for large scale cultivation of rice in the Barak Valley to gain out of the growing demand in Bangladesh market. It is true that rice is exported to Bangladesh from other parts of the country and through other borders, but the export of rice grown in Barak Valley through Karimganj will be more cost effective because of the shorter distance and lower transportation cost. The role of border trade in generating employment in Barak Valley is also no mean considering that even at the present level of trade at least 25,000 people are engaged directly or indirectly in the border trade.

Role of the Government

The Government of India, though belatedly, has come forward to promote border trade by extending the benefit of DEPB (Duty Exemption Pass Book) to traders and exporters of the Northeast Region (NER). The exporters of Barak Valley are also likely to be benefited from this incentive, provided there are enough products to be exported. The exporters of coal from NER were so long deprived of the benefit of transport subsidy, but recently the Government of India has taken a decision to extend this benefit to the exporters of this region. This is sure to benefit the exporters in Karimganj also. In fact, Karimganj is the only export center in the NER from where export can take place to Bangladesh through road, water and rail transport. The export from Karimganj is now taking place through water and road only, but it is bound to increase manifold if the old railway link between Bangladesh and Assam through Karimganj is revived. The railway track and custom stations are still there on both sides of the border The Indian railway has its track up to Mahisashan Railway Station in Karimganj district, while Bangladesh has its track from Shahbazpur Railway Station. Only 40 to 50 meters of track was withdrawn due to political reason. Before the partition (1947) there was a direct railway route between Sylhet and Kolkata via Mahisashan which would take less than thirty hours to reach Kolkata by the then low-speed meter gauge train. Therefore, if the old railway connectivity is revived all the exportable commodities, such as tea, sugar, ginger, rice, pineapple, automobile parts, building materials, etc from NER can be directly transported to Chittagong port and therefrom to the markets in South and Southeast Asia. The cost effective rail transport can help the revival of the Barak Valley's moribund tea industry, because Indian tea can be directly shipped to Pakistan and all other sea-bound countries which have markets for tea. The latest package of the Government of India of rupees three hundred crores on loan basis to boost the tea industry in Barak Valley can achieve the objective only if the profitable internationally marketing of the product can be assured by facilitating cost effective transportation. It should be accepted by all concerned that the revival of the tea industry of Barak Valley is possible by augmenting the productivity, ensuring quality and facilitating international marketing of the product, particularly in South Asia. If the owners of the tea gardens come forward with all their sincerity for making best use of the latest package and if the government facilitates the exportation through Karimganj, the lost golden days of the 'Cachar tea' can be brought to life. The management, employees and workers of tea gardens can celebrate *Durga Puja,* Dewali and *Fagua* once again in each garden with old gaiety and splendour.

The policy of the Government of India to enhance the country's international trade, the Prime Minister's emphasis on forging closer economic ties and trade relations with the countries of SAARC and ASEAN, and the much publicized 'Look East Policy' have opened up new opportunities of the border trade of the NER, particularly through Karimganj. It is by now apparent that the Government of India has been working for the development of export trade through Karimganj. The trade center at Sutarkandi has been formally inaugurated on 12th January 2007 by the Minister of State for Commerce, Shri. Jairam Ramesh, in presence of the Heavy Industry Minister, Shri Santosh Mohan Deb. Post office, Bank branch, offices of BSNL and the Customs Department, and the warehouses are likely to be established in the center shortly. Therefore, the role of the Government of India seems to be positive. However, this will not be enough unless the Government of Assam also comes forward to do its part in developing the internal roads and enhancing the agricultural and industrial productivity of Barak Valley.

Constraints-resolution

The Government of India is generally held responsible for the industrial and economic backwardness of Northeast India. This feeling of deprivation of the people of the region is natural and genuine when we compare NER with other regions of the country. Many attribute the apathy of the Central Government towards NER as one of the important causes of secessionist movements in the region, and these secessionist movements have come on the way of investments from both within and outside India. It is said that the investors are not inclined to risk their life and money in the 'terror stricken northeast'. However, there are many who believe that the 'terrorism' is cited only as a pretext to deprive the region of development. Moreover, the law and order has never been a problem in Barak Valley. Ironically, Assam, which is a victim of the so-called step motherly attitude of the Centre, has been extending the same treatment towards Barak Valley in respect of economic development of the valley. The general feeling in the Barak Valley is that the Government of Assam not only failed to play its role in economically developing Barak Valley at par with the Brahmaputra Valley due to its discriminatory policy but also failed to project the requirements of the Barak Valley in various plan proposals submitted to the Central Government. Thus, the attitude of the government of the state is considered as a serious constraint for the economic development of Barak Valley.

To resolve the problem of the State Government not taking adequate interest in the economic development of the Barak Valley, the public opinion in the valley is building-up in favour of the demand for a separate development council for Barak Valley. This demand has been placed before the Government of India by the Barak Upatyaka Bahu-Bhasik Samanway Samiti in August 2006. According to the demand, the 'Economic Development Council' should be like the Development Council for Bidharva region in Maharastra.

The Council should be composed of MPs and MLAs from the valley and experts in different fields to be nominated by the Central Government, one of the MPs from Barak Valley should be nominated by the Central Government as president of the Council, and fund should be allocated directly by the Centre to the Council for financing the development projects in the Barak Valley. The secretary of the Council shall be appointed by the Central Government and this officer should be equivalent to that of a Secretary to the Government of India. The Council should have a secretariat and a planning cell consisting of experts (government officials). The Council should also have an Advisory Committee consisting of the representatives of all the communities in the Barak Valley so that every part of the valley is developed evenly and the benefit of development is shared by all sections of people in the valley. The Council should come into existence through an Act of the Parliament, formulate need based micro and macro level plans and projects, and monitor the implementation after the proposals are sanctioned by the government. The demand for such a council was initially raised by the Barak Upatyaka Banga Sahitya-O-Sanskriti Sanmelan (a literary and cultural organization of the valley) which organized a convention of the representatives of all linguistic groups in the Barak Valley at Karimganj to discuss the problems of economic development of the valley and in that convention the Barak Upatyaka Bahu-Bhasik Samnway Samiti was formed to place the demand for the Economic Development Council. Ultimately, the Barak Upatyaka Bahu-Bhasik Samanway Samiti submitted the memorandum to the Prime Minister of India on 14th August 2006.

The response of the Government of India to the demand for economic development council is not yet known, but the fact remains that the demand is genuine, just and legitimate. If this proposal materializes then the scope for misunderstanding, discontent, distrust, ill-feeling, allegations and counter allegations within the state will be shrunken. In fact, the demand for the economic development council has been raised by various organizations and a central demand committee has also been formed. There should be joint and concerted effort from each and every organization who wants this Council to ensure accelerated economic development of the Barak Valley. The border trade will get a further boost under the patronage of the Economic Development Council.

Conclusion

The border trade offers an excellent opportunity for the reconstruction of the economy of the Barak Valley which was shattered by the partition of India in 1947 and the people of the valley and the governments at the centre and of the state should leave no stone unturned to reap the fruits of the globalization of economy by promoting trade between NER and Bangladesh. The spirited announcement of the Prime Minister of India at the SAARC conference in Delhi to allow goods from South Asian countries to take entry into India without customs duty has brightened the prospects of border trade. It is expected that this gesture of the Government of India will be reciprocated by other member countries of the SAARC sooner or later. In the meantime, the Union Commerce Ministry has demonstrated its sincerity in promoting border trade with Bangladesh through Karimganj by establishing the trade centre at Sutarkandi, and it is understood that the proposal for the revival of the old railway track is under consideration of the railway authorities. Therefore, it is time for the Government of Assam to take firm steps for drawing well thought out plan for developing infrastructure in order to ensure enhanced rate of border trade from Karimganj. This is indispensable for registering growth rate in the economy of Barak Valley and Assam, from the present level of 5% in case of the entire state, and perhaps less than that in case of Barak Valley, to the level of national average. This is also necessary in the interest of the people of the valley in the form of increased per capita income and improved standard of living. "Export or perish"—the slogan of the first Prime Minister of Independent India, Jawaharlal Nehru, is much more relevant today.

REFERENCES

Bhattacharjee, J B (2000), *Trade and Colony: The British Colonization of North East India*, NEIHA, Shillong.

Das, Gurudas & R K Purkayastha ed. (2000), *Border Trade – North East India and Neighbouring Countries*, Akansha Publishing House, New Delhi.

Indian Commerce Bulletin, vol. IX, A Journal of Research on Business Studies, Commerce Association of Kerala, Tiruvantapuram.

Information obtained from Customs Department, Karimganj, and from the exporters of Karimganj.

Memorandum submitted in 1997 to the then Prime Minister, Shri A B Bajpai, by the North East Federation of International Trade, Karimganj.

Smith, David A (ed) (2004), *A New World Order? Global Transformation in the Late Twentieth Century*, New York.

11

Exploring Human Resource for Sustainable Development of Barak Valley

—*Jayati Bhattacharjee*

Of all the resources in Barak Valley, the most abundant seem to be the human source.

As per 2001 census, this small geographical area has a total population of 29,88,797 souls (Cachar -14,42,141; Karimganj – 10,03,678; Hailakandi - 5,42,978). The decennial growth during 1991-2001 was 18.66 in Cachar, 21.35 in Karimganj and 20.92 in Hailakandi, while the density of population in the three districts in 2001 recorded 381, 555 and 408 respectively. The literacy figures are better than the national average : Cachar - 8,35,888 (Male – 4,81,106, Female - 3,54,782), Karimganj - 5,59,512 (Male – 3,17,584, Female - 2,41,928), and Hailakandi - 2,66,909 (Male - 1,57,474, Female – 1,09,435). The economic classification of population in the three districts together returned 9,50,229 persons (Male - 7,54,064, Female – 1,96,165) as 'total workers', and of them, 7,35,395 (Male - 6,35,663, Female - 99,732) as 'main workers' and 2,14,834 (Male – 1,18,402, Female - 96,432) as 'marginal workers', thereby leaving

20,38,568 persons (Male – 7,84,836, Female- 12,53,732) as 'non-workers'. There are, of course, 2,40,480 (Male - 2,12,015, Female - 28,365) 'cultivators', 1,28,021 (Male - 1,02,954, Female - 25,067) 'agricultural labourers', 34,859 (Male - 13,593, Female - 21,266) involved in 'household industries', and 5,46,949 (Male - 4,25,552, Female - 1,21,397) 'other workers'. The figures are quoted from *Basic Statistics of NER 2002* [NEC, Shillong, 2002.] The publication does not provide the data relating to unemployment, particularly educated unemployment in which we would be more interested. Nonetheless, this official publication of the North-Eastern Council (NEC) provides some startling clues to wanton waste of human resource in the Barak Valley.

If the figures are taken as correct, of the 29.89 lacs population in the valley only 9.5 lacs are 'workers' and 20.39 lacs are 'non-workers', inclusive of male and female. Since the total population has thus been divided into 'workers' and 'non-workers', we have to read 'workers' as 'employed' and 'non-workers' as 'unemployed'. The 'workers' would, therefore, include those who are employed in various service sectors (including white-color employments) and other trade and callings. Of the 9.5 lacs 'total workers', 7.35 lacs are 'main workers' and 2.15 lacs are 'marginal workers'. Again, among these 'workers' 2.4 lacs have been shown as 'cultivators', .1.28 lacs as 'agricultural labourers', 0.35 lacs as involved in 'household industries', and 5.47 lacs as 'other workers'. These figures take care of 9.5 lacs 'total workers' (7.35 lacs 'main workers' + 2.15 lacs 'marginal workers'}. Of the 20.39 lacs 'non-workers' or unemployed, 7.85 lacs are male and 12.54 lacs female. A clear case of gender inequality apart, these figures shows that more than two-third of the total population is unemployed or unutilized in the economy. Even if 50% of these 'non-workers' are children and old or infirm, the adults unemployed in the valley are at least 10.19 lacs (or more than one-third of the total population). If a portion of the marginally or partially employed category, i.e. 2.16 lacs (which is in any case under-employed) is to be added to this figure, it would indeed be considered as staggering by any standard. The sustainable development shall remain a distant dream if such a large section of the population has to remain unemployed or unproductive. Therefore, an appropriate human-power planning is necessary to utilize the human resource in the valley by ensuring work for everybody. If the human resource is underutilized, the economy is bound to remain underdeveloped. A sustainable alternative development strategy for Barak Valley is perhaps industrialization which can absorb the surplus labour in tea gardens and agriculture, besides augmenting the overall economic growth.

In the specific context of the Barak Valley, however, the more pressing cause of frustration and disappointment among the youth seems to be the mounting problem of educated unemployment. The bulk of the educated and trained human-power here is remaining unutilized, thereby causing a colossal waste of the indigenous human wealth. When we talk of the development strategies for Barak Valley, the strategy for utilizing the services of the educated youth for the development of the valley should get the priority in our discourse. This is necessary not only to provide them employment but also to ensure that this army of the educated youths is not misused by those who are promoting militancy and insurgency in other parts of the Northeast. [Bhattacharjee, 2007] It must be understood that a Central University with about forty affiliated colleges, a NIT, a Polytechnic and similar other institutions of learning are producing qualified and trained human power in thousands every year, whereas in the absence of industry and private sector the scope of employment for this important category is almost next to nothing. The situation urgently calls for a perpetual industry-institution reciprocal tying-up for absorption of the educated youths. Such collaboration has been looked upon for a long time for generation of resources for the institutions by providing research and development services to the industries, but we are now talking of similar collaboration for generation of employment for the students. To make such collaboration reciprocal and sustainable, the institutions have to ensure quality and offer courses to suit the requirements of the employers. It is a known fact that the brighter students and those from the affluent families in Northeast (including Barak Valley) are migrating to Bangalore, Pune, Delhi and other places after +2, not only for technical and professional studies but also for general degree courses. A survey conducted by the Institute of Northeast India Studies among those students in Bangalore in 2007 revealed that the employment after graduation is the major consideration for students' mobility. It is said that even an arts student in a college in Bangalore can expect to get a job through campus recruitment. When pointed out that general education in Bangalore was much more costly for parents than in the Northeast, the almost common reply was that 'it is better to get expensive education and employment together than cheap education and unemployment'. The common feeling in the Northeast is that in the absence of industry and private sector, there is hardly any possibility of such industry-institution tying-up. However, the example of Sikkim Manipal University, which has been able to tie-up with employers elsewhere outside the region, should be an eye opener. Moreover, some companies from outside are now organizing 'recruitment fairs' in various towns, particularly in Shillong and Guwahati, once or twice a year. The institutions in Silchar should also be able to attract those companies to recruit their products through campus interviews. However, the quality of education, choice of courses and proficiency in languages shall matter a lot.

Finally, it is necessary to insist on the fact that the population is never a burden on the economy. On the contrary, it is a resource for economic development. The concept of 'so many mouths to feed' must change, and the attention should move from mouth to the head and hands. After all, the development is for the people and it can be done only by the people. The ideas come from head and the hands do the work. What is needed is appropriate planning that can take into account the population in an area and its needs and potentials of growth. It is expected that NEC, which is the nodal agency of the Government of India for the economic development of the Northeast, should come forward to take positive steps towards human-power planning for optimum utilization of the indigenous human resource for economic growth of the Northeast as a region. The NEC has a directorate of human resource development and, therefore, it should not be difficult for the Council to undertake such an exercise in research and analysis by involving the academia and the expertise in the universities and institutions of higher learning in the region to enable the NEC to formulate short and long term plans of action. The constitutional and political heads of the constituent states, who are members of the NEC, may make use of the studies by the experts for their respective states. It is also possible that the NEC on its own part looks upon the human resource as a resource of the region and designs appropriate schemes for the development and utilization of this resource for the sustainable development of the Northeast. The NEC is also expected to accept the fact that the Barak Valley deserves special attention for the highest density of population in the entire Northeast region and as a distinct culture zone with a history of own that is different from the rest of the Northeast.

REFERENCES

Bhattacharjee, J B (2007), *Roots of Insurgency in Northeast India*, Akansha Publishing House, New Delhi.

Morrison, Berrie M (1980), *Political Centres and Cultural Regions in Early Bengal,* Rawat Publishing House, Delhi.

North-Eastern Council (2002), *Basic Statistics of NER 2002*, NEC Secretariat, Shillong.

12

Development of Barak Valley: Question of Sustainability

—*Abhik Gupta*

The word "development" can have several meanings. A widely perceived connotation of development in the context of a district, a state, a country or any other political or geographical unit relates to the nature and extent of exploitation of its resources. It could also mean infrastructural development in terms of road, rail, surface water or air connectivity, or that of various industries, irrigation networks, telecommunication, information technology, and so on. These in turn have bearing on and are also dependent upon the quantum of exploitation of natural (or even human) resources in that area. The ecological concept of development adds another dimension to this definition. It takes into account the growth and maturation of an ecosystem or a landscape over time, which, however, is deeply influenced by the pattern and magnitude of anthropogenic activities in that area. This essentially makes development a historical process spanning past, present and future generations, and brings into play the question of sustainability, which assumes greater significance each day as ongoing exploitation renders the resources increasingly scarce and economically as well as ecologically unviable to be extracted and utilized. The term 'sustainable development' came into limelight after the publication of the report of the World Commission on Environment and Development, more popularly known as the 'Bruntland Commission' after the name of its Chair, Gro Harlem Bruntland, in 1987. The Commission's definition of sustainable development as "development that meets the needs of the present without compromising the ability of future generations to meet their own needs" has since become the most oft-quoted definition of sustainable development. This definition also recognizes development as a historical process. Therefore, in our effort to chart a roadmap of development for the Barak Valley, it may be worthwhile to make an appraisal of both the constraints and prospects in a historical as well as a geographical perspective as the latter provides the setting for the former.

Resources of Barak Valley: a Brief Historical-Geographical Perspective

Barak Valley, located in the southern part of the Northeast Indian state of Assam, includes the three districts of Cachar, Karimganj and Hailakandi. The valley derives its name from River Barak, which is its main drainage. The Barak flows in an east-west direction through the valley. Its major tributaries include the Jiri, the Chiri, the Madhura and the Jatinga on the north bank, and the Sonai, the Katakhal, and the Dhaleswari on the south, flowing in a north-south and south-north orientation, respectively. The Barak splits into two branches, namely, Surma and Kusiyara that enter Bangladesh and rejoins to form the mighty Meghna. These two branches also receive some important tributaries within the Indian territory such as Rivers Harang (of Surma), Longai and Singla (of Kusiyara) besides several flowing from Tripura and Meghalaya to join either Kusiyara or Surma in Bangladesh. Thus the Barak Valley is a part of the Meghna Basin, the latter narrowing down in its northeastern extremity where it is hemmed in by the mountains of the Jaintia Hills and the Borail Range to the north, the Bhuban Range bordering Manipur to the east, Mizoram to the south, and that of Tripura to the south-west. Lying approximately between 24^0 8' – 25^0 8' N latitude and 92^0 15' – 93^0 15' E longitude, the Barak Valley covers a geographical area of *c* 6922 km^2, of which some 3839 km^2 (55.46 %) comprises forests that are mostly confined to the peripheral hill areas. The central plains abound in wetlands that occupy a total area of 13737.5 ha.

The geographical isolation of Barak Valley from the main Ganga-Meghna basin of Bengal played an important role in governing its developmental history that can be said to have begun in the nineteenth century when in 1832, the British annexed this area into its vast empire and threw it open to the Indian and even the global market. Prior to this period, although the British had already taken Sylhet (which included the present-day Karimganj district) in the downstream of Surma-Kusiyara under its control, they did not express much interest in this remote upstream tract in spite of repeated appeals for rescue from the beleaguered Dimasa kings who ruled this area since about 1750. The Burmese invasion of Cachar was perhaps instrumental in changing the British perception and when they finally moved into this valley, they initially did so in strategic interest. However, they soon realized the immense potential of this region in terms of agriculture and forestry, and a couple of decades later, for setting up tea plantations. The foothill forests were cleared to establish tea gardens and to meet the demands of this rapidly growing industry, timber and other products were also extracted without any restriction from the hill forests. From 1832 to 1862, the forests were first under the control of the Public Works Department, and then the revenue wing of the Deputy Commissioner. Although a Forest Department was set up in 1862, the first reserve forests in

Sylhet and Cachar were notified only in 1878. Fellers could operate without any restriction of girth size except some lower limits (1.22 m) for Jarul (*Lagerstoemia flosreginae*) and Nageswar (*Mesua ferrea*). Consequently, a large portion of all mature trees, especially those near the streams and rivers, were felled and marketed, along with wanton destruction of young trees, as noted by Gustav Mann, the then conservator of forests, Assam, in 1873-74. Not only timber, valuable non timber forest products (NTFP) such as cane, honey, bamboo, ivory, agar and others were harvested in a most unsystematic and unsustainable manner. The entire 'chunam' (lime) boat fleet in Sylhet for transporting limestone was constructed from Jarul trees harvested from Cachar forests. The soft wood of Cachar had its use in the packing industry of Calcutta; umbrella handles, bamboo mats, cane products and 'murta' mats were readily sold in the markets of East Bengal and Calcutta. Cachar bamboo fed the Titagarh Paper Mill with raw materials. In the low-lying valley areas, wetlands were cleared of the dense swamp forests for agricultural expansion. The rich wildlife inhabiting the forests were also adversely affected by habitat degradation as well as hunting for trade and sport. Such short-sighted policies guided by a 'frontier attitude' that views the earth's resources as inexhaustible continued to be pursued till very recently, with obvious disastrous consequences [Bhattacharjee, 1957; Bhattacharjee, 1977; Anonymous, 1991; Gupta, 2001].

In a geographical-ecological perspective, the topography of the Barak Valley terrain is of a highly undulating nature with the central alluvial flats frequently broken by isolated hillocks or 'tillahs' and small hilly tracts – mostly running in a north-south orientation -that merge with the outlying hill ranges. In such an intricate hill-valley complex, it is essential to protect the ecological integrity of the hills in order to maintain the productivity of the valley areas. Rampant deforestation in the mountains in the peripheries of Barak

Valley including those of Meghalaya, Manipur, Mizoram, Tripura and North Cachar Hills resulted in soil erosion and consequent silt deposition in rivers and wetlands, reducing their carrying capacity and productivity. These historical processes greatly affected the ecology of this area, setting forth a chain of events leading to progressive decimation of its natural resource capital and ecosystem services.

The Present Scenario: Constraints and Prospects

Barak Valley is neither endowed with mineral resources nor has ample hydroelectric power development potential. Although it has natural gas reserves and possibly oil deposits, the actual potentials are not assessed till date and it is not clear whether natural gas and/or oil extraction can become a major contributor to development in future. The main strength of the Valley, therefore, is its biological capital, including agricultural, forest, river and wetland (fishery) resources. However, as discussed in the preceding section, the bio-resource base of Barak Valley has been severely degraded due to a historical process of systematic abuse and unsustainable extraction without adopting any worthwhile measures for its conservation, restoration and regeneration. To make things worse, the Valley had to bear the brunt of the partition of Bengal, in the aftermath of which it had to accommodate a large number of displaced persons from the erstwhile East Pakistan. In the absence of planned rehabilitation policies and programmes, the already fragile bio-resource base of the area was further decimated to meet the demands of the burgeoning local populace. A brief assessment of the biological capital of Barak Valley is attempted in the following sections to understand the present status, the nature and extent of the constraints. and prospects, if any.

Forest Products

The partition of India in 1947 robbed Cachar of its

principal markets in East Bengal. However, even if it had existed, the extraction bonanza would not have lasted long. Despite a much truncated post-independence market, the scarcity of raw materials was felt as early as in the 1950s. Several forest-based industries had earlier existed in Cachar and Karimganj. The Surma Match and Industries Ltd., Karimganj, a flourishing concern supplying plywood to tea gardens of Assam and Duars; the K.S.C.M.C. Oil, Rice and Saw Mills, Madhurband, Silchar, working partly as a saw mill; the Bhanga Bazar Plywood Manufacturing Works, Bhanga Bazar; the Surma Valley Saw Mills Ltd., Bhanga Bazar; and the Hindustan Plywood Ltd., Badarpurghat, all started facing shortage of raw materials by the early 1950s and were unable to operate at their full capacities. The grossly unsustainable extraction of more than a hundred years without providing for regeneration had taken its toll. In the 1950s, when a modest enterprise was started in Patharkandi, Karimganj district, for distilling agar oil, the owner found it difficult to procure enough raw materials (Bhattacharjee, 1957). The Cachar Paper Mill at Panchgram, Hailakandi district, which is owned by the Hindustan Paper Corporation (H.P.C.), is another example of unsustainable planning without caring to learn lessons from history. Instead of making any serious effort to create a sustainable supply base by encouraging and supporting local farmers to go for planting bamboos and by raising bamboo plantations in wastelands, it solely depended on the forests of North Cachar Hills and Mizoram for meeting its demand. The Mill has ceased production in June, 2008, for want of bamboo and faces an uncertain future. However, it had succeeded in raising the price of bamboo in the local market, thereby hitting hard the rural poor who use it for house construction as well as the village craftsmen. Table 1 shows the forest statistics of the three districts of Barak Valley (FSI, 2005). Although the total forest area in Cachar and Hailakandi are around 58 % and that in Karimganj is around 46 % of the total geographical area, the proportion of very dense forest is extremely low, ranging from 0.24 % in Karimganj to 3.6 % in Cachar. The moderately dense forests are around 40 %, while the remaining 54-60 % is open, degraded forest. The forest department's revenue-oriented assessment of Cachar forests (Table 2) also portrays a dismal picture. While the open forests offer very poor prospects for forest product extraction, any further exploitation of the moderately dense forests would strike the proverbial 'last nail in the coffin' of the remaining wildlife habitats and seriously hamper the ecosystem services with disastrous ecological consequences. Nevertheless, even from these degraded forests, some 60 Non Timber Forest Products (NTFP) are still being harvested to provide livelihood to a large number of forest dwellers and traders. About 20 such items find their markets outside the state as well. However, the availability is fast declining and trying to build any development blueprint based on the forest sector without first achieving a robust regeneration of these ecosystems would comprise a foolhardy and futile exercise.

Tea – the flagship industry of Barak Valley – is beleaguered with many problems. A large number of tea gardens have either closed down or are seriously sick. Even those gardens that are doing relatively well, are struggling to maintain *status quo*. A fresh surge of development based on tea, thus, does not appear possible at the present time.

Agriculture

The rich and fertile soil of Cachar, especially in the floodplains of various rivers such as Sonai, Rukni and others had attracted the attention of Pemberton (1835) who was struck by the "... beautiful and fertile valley of the Barak". He described the land on the south bank of the Barak River as "...the finest plain, almost wholly unoccupied, through which the rivers Dullaseeree and Soonaee flow from south to north...". In fact, Sylhet and Cachar were often dubbed as the granaries of Bengal and Assam, for their bountiful production of paddy. The farmers of this area had a long experience with paddy and created a large number of local landraces with different properties. The soil hardly required any fertilizer supplement as every year the rivers deposited nutrient-rich alluvium on their floodplains, which also supported flourishing capture fisheries, especially in large floodplain lakes (locally called 'haors' or 'beels') like Chatla, Jabda and Lucca in Cachar, and Shonbeel and Ratabeel in Hailakandi and Karimganj districts. In the post-independence era, extensive flood control measures were adopted in the form of construction of embankments, the total length of which run into over 550 km. While these structures have indeed offered some protection from floods, they have also created numerous problems. The silt load carried by the rivers, which was earlier deposited in the floodplain, now gets deposited inside the river channel, not only resulting in elevation of the riverbed, but also depriving the floodplains of their natural nutrient subsidy. Furthermore, the composition of the sediment load during floods has changed over the years. While it was earlier dominated by the organic matter rich topsoil washed from the upper catchments, now it mainly comprises nutrient-poor coarse sand because of heavy soil erosion in the hills. The embankments do not have 'fish passes' and consequently act as barriers to the lateral migration of 'white fishes', which move into the inundated floodplains during the monsoon months for feeding and spawning. Several species of fishes like *chapila* and *bacha* (*Gadusia chapra* and *Eutropichthys vacha*, respectively) which once formed fisheries of considerable economic importance in the floodplain wetlands are now on a rapid decline. Indian major carps like *rohu* (*Labeo rohita*) and *katla* (*Catla catla*) are also facing decimation of their numbers due to a sharp reduction in their spawning areas [Gupta, 2003]. Large-scale settlement of displaced persons from erstwhile East Pakistan as well as the local increments has resulted in overexploitation of the fisheries and destruction of spawn, thereby preventing natural regeneration of fish stock in the rivers. Thus, in spite of Cachar and Karimganj having the largest area under seasonally inundated floodplains among all the districts of Assam, and a substantial number of oxbows (*anua*) (Table 3), the capture fishery potential has drastically declined in the recent years and the total catch is unable to meet even the local demand, let alone that of the outside markets. Prospects of agriculture in Barak Valley are also affected by progressive diminution of individual land-holding size, low consumption of chemical fertilizers and other agrochemicals and undeveloped irrigation system. In the absence of any irrigation facilities, the farmers face great difficulty in procuring water. Only those farmers who own plots located in the lowest elevations in the floodplain have easy access to water-filled depressions or small natural water channels that they can impound to obtain water. Farmers having plots in the middle and upper layers of the floodplain suffer from acute water scarcity. Pumping up water from distant sources is both costly and cumbersome and can be afforded by few. Thus a paradoxical situation exists where the lower plots have better access to water but are exposed to a greater risk of inundation by floods that would destroy the crop. In contrast, the middle and upper plots are relatively safe from floods, but are constrained by the scarcity of water in winter. The situation has worsened over the years as deforestation in the catchments of the feeder streams and rivers, especially in the hills, has led to soil erosion and deposition of silt in the watercourses that drain the floodplains. The floodplain lakes that used to retain water in the dry season have also become progressively shallower due to silt deposition and/or due to reclamation for agriculture. Taking advantage of the poverty of the floodplain farmers, brick kiln owners are grabbing the agricultural lands for making bricks, which now have a high demand because of the construction boom in the cities of Barak Valley. This demand is being met by gobbling up good agricultural land, cleverly bypassing legal restrictions. The brick-kiln owners lure the poor farmers into leasing out their land for three years and offer a yearly sum of around Rs. 20000 per hectare. When the land reverts back to its original owner, it is defaced and virtually useless. This is a very serious ecological issue that has its root in the declining fish and agricultural productivity in the floodplains and the resultant impoverishment and financial insecurity of the farmers [Gupta, 2006].

Wildlife

Wildlife, especially its animal components, can be a very important natural resource of any area, considering the surge of eco- or nature-tourism the world over. Being an integral part of the 'Indo-Burma Biodiversity Hotspot', which is one of the 25 such Biodiversity Hotspots all over the world [Myers *et al.*, 2000], the Barak Valley harboured high plant and animal diversity in its forests and wetlands, being no less richly endowed than the Brahmaputra Valley. Sadly enough, this once-rich diversity has dwindled over the years due to the factors outlined in the previous sections of this paper. To cite a few examples, two species of rhinoceros, namely the Javan and the Sumatran rhinoceros (*Rhinoceros sondaicus* and *Dicerorhinus sumatraensis*, respectively) that are different from the one (*Rhinoceros unicornis*) found in the Brahmaputra Valley were once found in the forests of Barak Valley and Sylhet [Wood, 1930; Milroy, 1935; Rookmaker, 1980]. Indiscriminate hunting and destruction of habitats have been responsible for the local extinction of these two prized species that could have attracted a large number of wildlife researchers and nature/ wildlife lovers from all over the world. The swamp deer (*Cervus duvauceli*), once abundant in the swamp forests of the wetlands has vanished from the Valley. The tiger (*Panthera tigris*) has also shared the same fate. During the late 1800s and early 1900s, ivory was obtained without any restriction from the forests of Sylhet and Cachar, resulting in selective hunting of male tuskers and endangering the elephant herds. Furthermore, deforestation and incursion of agricultural plots deep inside the forest caused shrinkage of elephant habitats and scarcity of forage for the pachyderms. Consequently, elephant depredation of field crops as well as damage and destruction of property and manslaughter led to severe man-elephant conflict that started in the pre-independence days to reach its climax in the 1990s. The elephant population dwindled from some 200-300 in the early 1960s to less than 30 at the turn of the century and perhaps even fewer today. It is now apprehended that the elephant populations of Cachar, Mizoram and Tripura can virtually be written off as doomed forever [Lahiri Choudhury, 1998]. The hoolock gibbon (*Hylobates hoolock*), which is the only Indian ape, is found only on the south bank of Brahmaputra, including Barak Valley. The ululating cry and antics of this animal - called '*ulluk*' in Bengali - had ensured it a place in many folklores, myths and proverbs in Sylhet and Cachar, indicating that it was indeed abundant in the forests of this area. Today, a few isolated bands have so far managed to withstand the pressures of habitat loss and hunting and unless protected pro-actively, are likely to go down the extinction path. The Spectacled or Phayre's Leaf Monkey (*Trachypithecus phyrei*) and the mountain goat 'serow' (*Capricornis sumatraensis*) among the mammals, and the hornbills, the wood duck, the pheasants, the Burmese peafowl (*Pavo muticus*), the Swamp Francolin, and others among the birds, are all a part of the unique wildlife diversity of this area that is confronted with the threat of extinction. The gharial (*Gavialis gangeticus*) has vanished from River Barak, although a few dolphins (*Platanista gangetica*) are still sighted. Four species of soft-shell turtles survive in R. Barak and its tributaries as well as in some wetlands, although regular selling of turtle meat in the markets spell impending doom for these endangered taxa that find their place in the IUCN list of endangered species [Gupta, 2001; Das and Gupta, 2004].

Tourism

Despite the occasional spurts of enthusiasm exhibited by the State Tourism Department, tourism has never really taken off in Barak Valley. In its present state, Barak Valley can hardly attract tourists as it does not have the typical tourist attractions found in many other parts of India. Even in terms of eco-tourism prospects, it lags far behind the Brahmaputra Valley, which not only has been able to conserve to a large extent its forests and wildlife in a chain of protected areas extending from Manas in lower Assam to Dipor Beel and Kaziranga in the middle and Dibru-Saikhowa in upper Assam. The Brahmaputra is itself a major attraction. Further, the Brahmaputra Valley is the gateway to tourist havens like Meghalaya and Arunachal Pradesh. Nevertheless, the Barak Valley could still aspire to find a place on the eco-tourism map if it can put an immediate halt to any further degradation of its forests and wetlands, and ideally undertake some eco-restoration activities. Several activities like river cruises along the upper reaches of Barak and Katakhal rivers, nature walks in the forests and tea gardens, trekking in the forests, etc., if promoted and publicized, could still attract visitors, even from abroad. There is a need for active cooperation with Mizoram, North Cachar Hills and even Manipur to offer a total package to the prospective tourist. This requires improvement of infrastructure in the form of restoration of the existing 'dak bungalows' and forest lodges and creation of new facilities wherever necessary. The internal road and even rail networks need to be repaired, improved and augmented. The folk dances and other art forms of the different ethnic groups, if revived and marketed properly, could prove to be yet another attraction. However, all these need imaginative planning, sincere political will, efficient administrative machinery and informed and pro-active civil society.

Conclusions

The political upheaval in the form of partition of India (and Bengal) had played havoc with the Barak Valley by robbing it of its market, its natural road, rail and riverine routes, its access to the seafront, and even assaulting its unique identity and self-esteem. Unsustainable exploitation has stripped it to a large extent of its natural resource capital. Hard reality, therefore, dictates that the development of Barak Valley has to be charted keeping in mind certain limiting, boundary conditions. Nevertheless, in the event of Bangladesh and Myanmar adopting more globalized and open policies, the scenario could drastically improve, although in the present state of affairs in both the countries, this remains a distant dream. Under the present conditions, a few strategies may be explored:

1. Quality- rather than quantity-oriented approaches in agriculture, where certain paddy landraces could be revived and marketed abroad. These include fragrant cultivars like 'kalijira', 'sahebshail', etc. The high starch 'biroin' varieties could also possibly find markets in South East and East Asian countries where such rice is preferred. Rice sweets and rice cakes are very popular in South East Asian countries. Relevant agricultural research needs to be directed to create the necessary knowledge base and improved production technology.

2. Marketing of medicinal products like *chalmugra* oil, essence-yielding species like agar and '*gondhi kochu*', which are still being sold as NTFP. Both in-situ and ex-situ conservation and propagation is necessary.

3. Deer (*Muntiacus muntjac* – barking deer or chhagli horin) and turtle farming. Captive breeding

techniques need to be evolved and standardized. Turtle meat and carapace have high demand in South East Asian and Chinese markets. Deer farming is flourishing in New Zealand, while Taiwan and Vietnam have taken up turtle farming with a fair degree of success.

4. Horticulture development constitutes another avenue. Besides better-known fruits like pineapple and orange, the various indigenous fruits like some berries, citrus varieties like 'satkora' and others may be explored. Thailand has very successfully introduced its unique fruits to the western tourists. The jackfruit could possibly be marketed like the *durian* of South East Asia. Durian-flavoured chewing gums and boiled sweets are fairly popular in Vietnam. Wine-making from some of these fruits may also be explored.

5. Eel, catfish, small shrimp and even edible snail culture can have prospects in South East and East Asian markets. The small shrimp species of Barak Valley have a characteristic flavour and could be marketed as shrimp paste, especially in South East and East Asian markets.

Appropriate value-additions and processing could improve the marketability of these products. However, the above list is essentially illustrative and not exhaustive. The intention is to stimulate 'outside the box' thinking among planners and entrepreneurs. Entrepreneurs in Barak Valley could learn a few lessons from their counterparts in South East and East Asian countries like Thailand, Singapore, Malaysia, South Korea and Vietnam, which have succeeded in aggressively marketing their indigenous products to novelty-seeking, health-conscious consumers from the west.

A popular perception has grown in Barak Valley that the implementation of broad gauge railway and some mega-projects like the Tipaimukh dam can work miracles for developing this region. The broad gauge will undeniably relieve thousands of commuters from their present misery. However, without a rejuvenated resource base, there will be inflow of goods from outside and outflow of Valley people's money, without any long-term contribution to the development of the Valley. Similarly, augmented power supply would simply illuminate homes and not factories and industries. It is to be realized that improved communication and power supply are mere tools for development. They are like enzymes that can catalyze development, but essentially need the substrate of a sustainable natural resource base to act upon in order to bring about all-round, long-lasting development. While devising development strategies for Barak Valley, the issue of sustainability ought to be given the attention that it has not received so far.

Table 1
Forest Statistics in Barak Valley

District	*Geographical Area (km²)*	*Forest Area in km² (% of Total Forest Area)*				*% of Geographical Area*
		Very Dense	*Moderately Dense*	*Open*	*Total*	
Cachar	3786	81 (3.6)	929 (41.8)	1215 (54.6)	2225	58.8
Hailakandi	1327	13 (1.7)	342 (43.8)	426 (54.5)	781	58.9
Karimganj	1809	02 (0.2)	323 (38.8)	508 (61.0)	833	46.1

Source: FSI, 2005.

Table 2
Forest Assessment of Cachar district, Assam

Forest Category	*% Distribution*
Well-Stocked	0.32
Medium-Stocked	2.33
Poorly Stocked	51.14
Mixed Bamboo	31.6
Pure Bamboo	1.87
Encroachment and Shifting Cultivation	6.03
Others	6.71

Source: Anonymous, 1991: Gupta, 2001.

Table 3
Natural wetlands in Barak Valley districts of Cachar, Karimganj and Hailakandi, Assam

District/ Region	*Wetland type*	*Number*	*Area (ha)*
Cachar	Lake/Pond	57	1151.5
	Ox-bow Lake/Cut-off Meander	29	592.5
	Seasonally Waterlogged	231	4869.5
	Swamp/Marsh	19	564.5
	Total	336	7178.0
Karimganj	Lake/Pond	6	95.0
	Ox-bow Lake/Cut-off Meander	1	87.5
	Seasonally Waterlogged	53	4667.0
	Swamp/Marsh	10	870.0
	Total	70	5719.5
Hailakandi	Lake/Pond	7	322.5
	Ox-bow Lake/Cut-off Meander	4	37.5
	Seasonally Waterlogged	36	480.0
	Total	47	840.0
Barak Valley	Grand Total	453	13737.5

Source: Space Applications Centre (ISRO), Ahmedabad, India (2000).

REFERENCES

Anonymous (1991), *Working Plan for the Forest Reserves of Silchar Division, Assam*. Forest Department, Govt. of Assam.

Bhattacharjee, J. B. (1977), *Cachar Under British Rule in North East India*, Radiant Publishers, New Delhi.

Bhattacharjee, P.N. (1957), *Working Plan for the Forest Reserves of Cachar Division, Assam*, Forest Department, Govt. of Assam.

Das, K.C. and Gupta, A. (2004), 'Turtle Market Surveys in Silchar, Assam, Northeast India'. *Turtle and Tortoise Newsletter* 8: 16-17. *Chelonian Research Foundation, USA*.

Gupta, A. (2001), Forests of Barak Valley, Assam: History, Management and Status as Wildlife Habitats. In: *Sustainable Management of Forests – India* (Eds. A. Arunachalam and M.L. Khan), pp. 197-212. International Book Distributors, Dehra Dun.

Gupta, A. (2003), Flood and Floodplain Management in North East India: an Ecological Perspective. In: *Proceedings of the 1st International Conference on Hydrology and Water Resources in Asia Pacific Region* (Eds. K. Takara, and T. Kojima), Vol. 1, pp. 231-236. Asia Pacific Association of Hydrology and Water Resources, Japan.

Gupta, A. (2006), Water Availability, Poverty and Socio-Economic Crisis in the Floodplains of Barak Valley, Assam, North East India. https://www.mpl.ird.fr/ UR044/projets/Textes/Gupta.doc.

Lahiri Choudhury, D.K. (1998), 'We belong to this earth, we also want to live'. *Environ* VI (2): 58-59.

Milroy, A.J.W. (1935), 'The Preservation of Wild Life in India, nr.3. Assam'. *J. Bombay Nat. Hist. Soc.* 37 (1, Suppl.): 97-104.

Pemberton, R.B. (1835), *The Eastern Frontier of India*. Mittal Publications (2000)., New Delhi, India.

Rookmaker, L.C. (1980), 'The Distribution of Rhinoceros in Eastern India, Bangladesh, China, and the Indo-Chinese Region'. *Zool. Anz.* 205: 253-268.

Wood, H.S. (1930), 'Observations on Indian Rhino and their Shikar on Foot'. *J. Darjeeling Nat. Hist. Soc.* 4: 59-69.

13

Quest for Development Strategies for Barak Valley

—Bibhash Ranjan Choudhury

The Barak Valley (as it is called today), now consisting of three districts (Cachar, Karimganj and Hailakandi) was called Cachar District (then consisting of Silchar, Hailakandi and Haflong subdivisions} when it was coalesced in 1874 with a British Province, called Assam, with which it had little cultural, social and economic affinity. It was the dawn of independence, following the partition of the country in 1947, that saw the coalescence of Cachar and Karimganj (earlier a part of Sylhet) with Assam, a major State of India in the Northeast.The partition rendered the Barak Valley completely landlocked and subjected it to some such socio-economic problems which were not to be easily resolved. These problems still persist and constraint the development of the valley. The valley constitutes a central part of the Northeast Region although characteristically it does not resemble much with the other parts of the Northeast - geographically, culturally or linguistically. Quite understandably, any discussion on the development strategy, therefore, calls for a brief discussion on the region itself.

The Government of India brought about a reorganization of the states in the country in mid-fifties of the last century when some new states came into existence by way of bifurcations of some of the larger states. This reorganization was effected mostly in accordance with the recommendations of the States Reorganization Commission (1954). Thereafter, there have been several instances of territorial readjustments between states, upgradation of union territories and centrally administered areas into states, creation of new states, and autonomous areas within states. In the Northeast of the country, there have been prolonged political, linguistic, ethnic and communal unrests which eventually culminated in the emergence of new states from within the geographical framework of the erstwhile British Assam. The states of Nagaland, Meghalaya and Mizoram are born out of Assam. The District Councils in Karbi Anglong and the North Cachar Hills have been more empowered, and the Bodoland Territorial Council BTC) has been created. The North-East Frontier Agency (NEFA), which was once under administrative charge of the Governor of Assam, is now the State of Arunachal Pradesh. One thing that was common in all the cases of turmoil preceding the emergence of these states through the reorganization of Assam was a feeling of insecurity amongst the minority people as to their ethnic and linguistic identity and a feeling of deprivation. The states like Meghalaya, Mizoram and Nagaland are clearly the products of this feeling of deprivation and a feeling of seclusion from the mainstream of the state of Assam.

That the new states like Nagaland, Mizoram and Meghalaya, which broke away from the fold of Assam, have substantially gained by the separation from Assam is borne out by the fact that the growth of Per Capita Net State Domestic Product in these states has been much faster in relation to the corresponding growth rate of Per Capita Net State Domestic Product in Assam in current prices during

1981-2001. During this period, there has been a tremendous growth of population in the North-Eastern States, yet there has been 10 to 15 times increase in Per Capita Net Domestic Product in these newly-emerged states, whereas Assam is lagging far behind them. If the statistical data generated by the Government of India are to be relied upon, one has to admit that the propriety of conferment of statehood to some areas of the erstwhile state of Assam has been vindicated by this faster growth of per capita state domestic product of the newly created states. There has been a visible transformation of the overall economic and social scenario of the people living in those states. The reasons for this positive change may be attributed to several factors, and some of these factors must unfailingly be the better exploitation of local potentials and a rejuvenated spirit of the people emanating from a feeling of an end of exploitation and deprivation, on the one hand, and on the other, from the satisfaction and self-respect arising from the conditions created for the conservation of ethnic identity. The experience of some of these states has inspired many other ethnic and linguistic minority areas within the state of Assam to demand autonomy or statehood for the respective sub-regions or areas. It is being increasingly believed that such a measure may pave the way for the development of the backward areas and their people. Therefore, the question on which the social scientists would have to ponder for quite a long time is: Do small territorial administrative units pave the way for better exploitation of development potentials with more involvement of people in the process? In the Northeast the answer to this question tends to be affirmative. Attainment of statehood by some territorial segments of the multi-ethnic and multi-cultural state of Assam after the independence of India has certainly stepped up the pace of progress in the new states.

The Barak Valley of Assam being one of such relatively backward areas, there is a feeling of deprivation among the people for a long time. It is, therefore, believed in some quarters that if the area is converted into a centrally administered territory or is given local autonomy, the development of the valley would be easier and faster. The question as to 'given the autonomy or statehood, how would the valley prosper' remains unanswered. In fact, there is no simple solution to the problem of economic backwardness of an isolated and landlocked geographical area like the Barak Valley. Industrial development alone cannot transform the economy of the valley, nor can agriculture alone bring about the desired change. The development should always mean a balanced and all-round development. The positive experience of statehood in some areas may not repeat itself in all other areas unless the conditions are also similar. Moreover, a micro-region like the Barak Valley might not be able to develop itself in isolation and, therefore, may call for a macro-economic approach for development.

It is, however, necessary to appreciate that the Barak Valley, a region recognized as one of the bio-diversity hotspots of the world, is a part of Assam and the Northeast but physiographically this valley is not altogether alike the other parts of the region. Despite being endowed with various development potentials, the economic development of Barak Valley has ever remained a hard-to-attain matter. It all began with the partition of India that rendered the valley cut-off economically, culturally and geographically from places on which its development depended. However, it is not the partition alone that has ailed the valley. If the partition was responsible for giving a serious jolt to the economy, it was, afterwards, distinctly the failure of the government to deal with post-partition problems of the valley which stood on the way of its development. The partition was also followed by the waves of exodus from the neighboring country, namely, East Pakistan (now Bangladesh) and, consequently, the valley had to accommodate a large number of displaced persons.

The problem was grave enough and it was then necessary to explore ways and means to rehabilitate not only the displaced persons but also to rehabilitate the economy of the valley which was shattered by the partition. It was exactly this that happened in the areas bordering West Pakistan, but in case of Barak Valley and other areas bordering East Pakistan that approach was missing.

It is said that there is nothing wrong in being a bit wishful when the development is wistfully desired. In Barak Valley, labourers engaged with tea-manufacturing accounted for 90% of factory laborers till the end of 1970s. The situation has changed a little thereafter despite the growth of a few industries. There is, thus, definitely a tremendous agricultural bias in the economy. So, considering this aspect and the existent infrastructural constraint to industrial development, and the extra-economic factors that play a serious role in this part of the country, the people are constrained, at times, to indulge in a wishful thinking that agriculture may be instrumental in the development of the valley. The valley has thus been perpetually eluded by any concerted effort or sincere desire and plan for comprehensive development and for remaking or rebuilding of the economy of the valley. In last six decades the people of the valley have almost been pushed to the wall, and when they are now in a desperate bid looking for a pathway or pathways to get out of poverty the agriculture seems to be the only potential sector that can be instrumental in the development of the valley. This wishful thinking is, however, not unfounded altogether. There are many success stories of agriculture as an engine of growth early in the development process and of agriculture as a major force for poverty reduction. Most recently, China's rapid growth in agriculture supported by liberalization of market and rapid technological change has largely been responsible for the decline in rural poverty from 53% in 1981 to 8% in 2001. Agricultural revolution predated industrial revolutions that spread from England in the 18th century to Japan in the late 19th century. In the 21st century too, agriculture continues to be a fundamental instrument for sustainable development and poverty reduction. Three of every four persons in developing countries live in rural areas and the most of them depend on agriculture for their livelihood. But agriculture has its limitations. Development of agriculture can initiate the growth of a laggard economy but agriculture alone cannot be enough to massively reduce poverty. In modern times and in the context of complex consumption pattern, agriculture can be a component of effective development strategy but its development alone cannot make sustainable development possible.

The plight of the laggard rural economy of the Barak Valley is further saddened by the presence of a huge volume of workless semi-skilled or unskilled labourers. A qualitative change in the rural economy can hardly take place until and unless these labourers are put in some employment. But the demand for labourers, even for semi-skilled or unskilled, will increase within a dynamic rural economy in both agriculture and the non-farm sector. This dynamic rural sector can, however, be a real proposition in the valley if there is an improved investment climate which can be induced only by the government through a package of measures like securing of property rights, improvement of inter-village connectivity, and other infrastructure, development of innovative approaches to credit and financial services and coordination of public and private sectors to encourage agro-based industry clusters. The measures indicated here are sure to provide a strong impetus for development. And it is the government that will have to initiate it. The government will have to have the real zeal and willingness to do it and sustain it. Any governmental effort, without being backed by the zeal and real willingness, is bound to prove futile in the long run.

There is also the need for taking care of the people around or places around. The development of the valley independent of the development of people or places around it cannot be thought of. It is here that the necessity of developing other sub-regions to facilitate the development of the valley becomes imperative.

It should be clear that agriculture alone cannot achieve the desired development for the Barak Valley and its neighbourhood. Agriculture can only initiate the transformation by enhancing income at rural sector creating thereby the necessary conditions for non-agricultural development. A people-based approach to development would require separate strategies for hills and plains. It has been estimated by National Institute of Public Finance and Policy that if the Northeast is to reach the all-India level of Per-capita Income by 2020 (estimated at about Rs. 78,000/- at 2006-07 prices), GSPD in N.E. States would have to grow at 11.8% per year. The achievement of this growth rate will call for the enhancement of 'capabilities', implying increase in productivity. This, once achieved, will enhance income and wellbeing and subsequently, the conditions to accelerate development would be improved.

It can hardly be denied that the chronic backwardness of the Barak Valley has rendered people, in general, less responsive to any dynamic change in the pattern of economic activities. On the other hand, human resources can play a very crucial role in development, if these are geared up and made agile to appropriate every opportunity for development that comes their way. Therefore, all out efforts are necessary to promote solidarity amidst diversity. People are to be accommodative and tolerant. This is necessary to create mutual trust - a factor necessary for encouraging investment. The otherwise of this will simply turn not only the valley but the whole of the Northeast Region into the caldron of ethnic conflicts making the real development of the region, thereby, an impossible proposition.

In other words, the strategy for development of Barak Valley should take into account the overall development of the Northeast as a whole. The sub-regions within the Northeast are inter-connected not only by geography but also by several other characteristic features. There is also a long history of living together and exchange of goods, ideas and services, despite ethnic and cultural diversities. The resources and markets are spread all over the region, the surface communication network for any one of the sub-regions must pass through several other sub-regions, and to contain flood in an area the river that causes flood must be tamed elsewhere. The Barak Valley is so located that its economy cannot be developed without developing the neighbouring areas like North Cachar Hills, Jaintia Hills, Manipur, Mizoram and Tripura which surround the valley. Therefore, the sub-regions within the larger region would have to be bound together by economic relations and this relation of interdependence would pave the way for a sustainable development.

Finally, none of the sub-regions in the Northeast is in a position to grow independently. An attempt on the part of any sub-region to develop itself independent of other sub-regions may for sometime generate activities, income, and employment, but these would not be sustainable for long due to the inadequacy of favourable internal market conditions that can accelerate the activities uninterruptedly. On the other hand, if a strategy of development based on sub-regional mutual help is drawn up the people will feel their mutual inter-dependence which would in the long run eliminate mutual conflicts creating thereby a congenial atmosphere for development. This will also call for a strategy where development will take place through people's participation, that is participatory development or what is nowadays called ethno-development. The Barak Valley, therefore, cannot dream of developing herself independent of sub-regions around her in the Northeast. However, in order to develop an appropriate strategy a sub-region must have the liberty to determine its own course of action. Otherwise, it will remain an ignored and backward area in the larger landscape of the Northeast. Similarly, other laggard areas will have to be given the scope to decide their own course of action. And this course of action would have to be determined within a given guideline laid down on judicious and realistic considerations, without being biased in any way. It should also be said that such a pathway of development can be followed only if there is a consensus on the strategies of development among the sub-regions or states in the Northeast. This will necessarily call for sub-regional or inter-state cooperation, coordination, agreement or even agglomeration. There may even be the formation of a consortium consisting of sub-regions as members. There may even be sub-regional trade agreement arrived at for mutual benefit. Thus the involvement of all, not only in the Barak Valley, but all the ethnic groups or all the sub-regions of the Northeast, will have to be ensured for the development of all. That is, there should be the development of all the sub-regions in unison, or there would be no development proper of any of the sub-regions. The strategy for the development of Barak Valley should, therefore, be a strategy of development with others so that the development, if achieved, is sustainable.

REFERENCES

Academic Foundation (2006), *Migration, Development and Poverty Reduction in Asia*, New Delhi.

Govt of India, Economic Survey Report – 2006-07, 2007-08, New Delhi.

Oommen, T.K. (1998), "Changing Paradigm of Development: The Evolving Participatory", *Journal of Social and Economic Development*, Vol. XI, No 2.

Rastogi, P.N. (2004), *Ethnic Tensions in Indian Society*, Delhi.

Sing, J.P. (2001), *Human Resources in North East India*, New Delhi.

UNDP (2008), *Human Development Report – 2007-08*, Fighting Climate Change.

World Bank (2008), *Agriculture for Development*: World Development Report, 2008.

14

Demand for Barak Economic Development Council (BEDC)

—Tarun Das

I

Regional inequalities call for specific intervention when they constitute problems. The justification for regional policy rests on the imperfection of market mechanism operating in a spatial framework. The Barak Valley as a distinct geographical and cultural space in the state of Assam requires special treatment at analytical and policy spheres. The state of Assam consisting of 23 districts can be broadly divided into three zones, namely, (i) the Brahmaputra Valley, (ii) the area comprising two hill districts of North Cachar Hills and Karbi Anglong, and (iii) the Barak Valley consisting of the districts of Cachar, Hailakandi and Karimganj. These three regions of Assam are geographically, historically, socially and economically different from each other in many respects. The Barak Valley has all the characteristics of a problem region.

The most tragic feature is that the entire Barak Valley has remained very poor in economic infrastructure. The valley, which forms the southern extremity of the state of

Assam, is surrounded by hills on three sides and has a riverine international border with Bangladesh on the other side. Before independence and partition of India, it was a part of the larger Surma Valley famed for its tea industry that came up under British colonial rule and the active response to the historic anti-colonial struggle for India's independence. But at the time of independence the Sylhet district of Surma Valley was combined with East Bengal to form East Pakistan, which was to become Bangladesh later on. As a result, the present Barak Valley, or the part of Surma Valley that remained in India, lost its vital road and waterway links with the rest of India (which were traditionally through East Bengal). Though the rail and road connections with the Brahmaputra Valley in the north is now in existence, the communication of the region with the outside world through these mountainous routes has always been difficult and prone to disruptions. Consequently, the movement of men and materials to and from the Barak Valley region has become costly.

The traditional tea industry has been on decline since independence and there was hardly any serious initiative for setting up new industrial units. The compound growth of tea industry in Barak Valley, Assam and all-India shows more or less a stagnant one. However, there has been a significant development in the tea-growing region of upper Assam in the form of emergence of small tea growers. Traditionally, tea plantation in Assam has been a monopoly of the large estates. But with the emergence of small tea growers, many households and co-operatives have started producing green leafs, which has significantly contributed in the growth of tea production in recent years. Such a trend is absent in the Barak Valley region. The relatively lower yield of tea in the region and the recent problems of diseases and pest encountered by the planters in the region might work to prevent the emergence of small tea growers in Barak Valley.

Nonetheless, tea plantation continues to be an important economic activity providing employment to 10.4 percent of the workforce of the valley.

The Barak Valley's backwardness in manufacturing activities can be understood from the fact that the household and non-household industrial units engaged in manufacturing, processing, servicing and repairing employ only 6.0 percent of the total workforce (as per 2001 census). The most of the industrial units outside the household sector are either agriculture or forest based. In private factory sector, the industrial units of the region are tea manufacturing units, plywood and saw mills. The only notable public sector unit set up in the region is the Panchgram Paper Mill under Hindustan Paper Corporation in Hailakandi district. The Cachar Sugar Mill in Karimganj district died a premature death. The long tradition of the region in handicrafts, weaving, carpentry, boat-making and manufacturing a variety of consumer goods, and the availability of cheap skilled and semi-skilled labour have not been utilized even for the growth of small industrial units, despite the prevalence of a favourable law and order situation that offered a more favourable environment than any other part of the Northeast for industrialization. The geographical isolation of the valley and the fragility of the existing infrastructure (particularly power, transport and communication) have been deterrent to the economic development of the region. The prospects of economic development through utilization and exploitation of growth potential of agriculture and development of agro-processing industries also remain a distant dream due to lack of initiative on the part of the authorities. The recurring floods are a cause of human misery that the people of the valley are doomed to silently suffer every year.

In respect of agriculture the region lags far behind the achievement at the all-India average level, although the yield rates in Barak Valley compares favourably with those of the state of Assam as a whole. In respect of the use of yield increasing inputs such as HYVs and fertilisers also a similar picture of underutilization looms large. In extents of adoption of HYVs and application of chemical fertiliser per hectare it still lags far behind the level of achievement in the country as a whole. Perhaps it will be more appropriate to evaluate the performance of agriculture in Barak Valley in light of the state of agricultural infrastructure – especially irrigation in the region vis-a-vis the same in the state and the country as a whole. In parts of Brahmaputra Valley of the state and in many parts of the country, the cultivation of HYVs in the rabi season has recently been acting as the prime source of overall growth in the production and yield of rice. Due to practically non-existent irrigation facility, the farmers in the Barak Valley region are not in a position to use the HYV-fertiliser package extensively in the rabi season. Thus the farmers have to use the seed-fertiliser technology mainly under rain-fed conditions in which application of fertiliser entails greater risk than under controlled watering by irrigation. In light of these infrastructure related constraints, the farmers of the region appear to have done well in not only putting a greater proportion of their rice acreage under HYVs than by the farmers in Assam taken together but also by applying chemical fertilisers at a significantly higher dose. The farmers in Barak Valley are thus clearly constrained by the poor state of irrigation infrastructure.

II

The people of Barak Valley have suffered economic backwardness for so many years since the Independence of the country. They have been mere spectators to the progress and prosperity that the rest of the country experienced in those glories sixty years. The people of the valley, with an economy that was shattered by the partition of the country in 1947, are now convinced that the initiative for the development of the valley must come from the people themselves. Therefore, it is proposed that the Barak Economic Development Council (BEDC) is constituted to act as a catalyst in economic development of the Barak Valley. As regards the organization and the aims and objectives of the Council, it has been proposed that:

1. It should have both popular representation and administrative wing consisting of experts in various fields. It must have command over the fund meant for the development of the three districts included within Barak Valley. The North-Eastern Council, the Central Government and the State Government of Assam may delegate authority and devolve fund to this body under statutory provisions. It must have command over the fund meant for the development of the three districts included within Barak Valley.

2. The Development Council should draw comprehensive plans for agricultural, industrial, infrastructural and human development of this region.

3. BEDC should be an effective development catalyst acting as a link between different sectors towards common objectives. It should provide a common platform for joint community efforts towards economic development of this region.

4. BEDC should have an effective linkage for a continual dialogue with the State and the Union Governments to solve various economic and developmental problems.

5. BEDC should conduct result-oriented seminars and conferences on current economic issues for the region. It should undertake special research projects and bring out studies and reports on various aspects of the economy in general and trade and industry in

particular. It should also provide detail analysis and free guidance to the entrepreneurs.

6. BEDC should provide up-to-date economic intelligence in respect of the Barak Valley region and guide the entrepreneurs, industrialists and Non-Resident Indians desirous of setting up or expanding their industrial units in the region.

7. It should publish regularly documents containing a snap-shop view of the latest economic development, useful data and spotlight on the region's economy.

8. A development council established for the Barak Valley can also work towards inter-state cooperation. Being located in the junction of Assam, Manipur, Mizoram, Tripura and Meghalaya, the Barak Valley has a special role to play in the economic development of the Northeast. A meaningful planned effort towards development of transport, irrigation and flood control, forest development and ecology protection, banking and trade in the Barak Valley will have regional impact spilling over the neighbouring states. Thus, one may envisage a much wider economic role of the Barak Economic Development Council in the development of Northeast India.

III

The rationale of the demand for setting-up the Barak Economic Development Council lies in the fact that a region that was economically totally devastated by one of the greatest tragedies in the history of the humankind, called the' Partition of India', deserved special schemes for the reconstruction of its shattered economy. It is a well known fact that this did not happen in the case of Barak Valley; the governments at the centre and of the state have so far totally failed to identify the need for a special dispensation for a very special region. Secondly, the Barak Valley is a landlocked low-lying area that is surrounded by the hills on three sides and also an international boundary on the fourth, and the lines of its communication with the rest of the state and the rest of the country through the mountainous tracts are too fragile and too vulnerable to ensure the regular movement of people and supply of essential commodities. Thirdly, although it is a part of Assam ever since the then colonial administration had tagged the area to the British Province of Assam, the Barak Valley is geographically, historically, culturally and linguistically a distinct region within Assam. Fourthly, economically it is the most backward part of the state of Assam. Its potentials of agricultural, industrial and human growth have remained unrealized due to the indifferent and apathetic attitude on the part of the government. Finally, there are so many instances in the country that the autonomous councils have been constituted for the development of the backward and minority areas within the states to empower the people to be the leaders of their own progress.

INTERACTIONS

A Discourse on Development Strategy for Barak Valley

—Sarbani Bhattacharjee

The Brainstorming Session on 'Development Strategies for Barak Valley: Constraints and Prospects' at Assam University, Silchar, on 29th April 2008, shall indeed remain memorable to all the participants for the extraordinary concern of the intellectuals in the valley to analyse the constraints and prospects of development processes on the basis of empirical evidence and to suggest appropriate strategies of development for the future The one-day academic discourse, organized by the Department of Economics, Assam University, in collaboration with Institute of Northeast India Studies, Kolkata, was divided into four sessions, viz. (i) inauguration, (ii) presentation, (iii) interaction, and (iv) valediction. The continuity of tenor between the sessions was the hallmark that made the discourse singularly absorbing to its theme.

Inaugural Session, chaired by Professor Tapodhir Bhattacharjee, Vice-Chancellor, Assam University, set the rhyme for the entire discourse. Dr Niranjan Roy, Organizing Secretary of the Brainstorming Session, in his welcome address, explained the objective of the discourse as an effort of the academia to contribute its share to development of the area and urged upon the participants for dispassionate deliberations on the issues at stake in the interest of the region and the country. The keynote address, delivered by Professor J B Bhattacharjee, Chairman, Institute of Northeast India Studies, delved into the historical sequence of development in the decolonization phase and briefly outlined the strategies for further analyses by the participating scholars. In his presidential remarks, Professor Tapodhir Bhattacharjee emphasized the need of understanding the problems of Barak Valley historically for identification of the constraints of development and insisted that the strategies proposed should take into account the aspirations of the people of the valley and its potentials of development. Elaborating why one has to go to history to understand the problems of development in Barak Valley and how did the partition of India affect the fortune of the people of the valley, the Vice-Chancellor was particularly emphatic about the role the Department of History, Assam University, is expected to play in bringing into focus the genesis of economic backwardness of the area and thereby strengthen the efforts of the University to fulfill its commitment to contribute positively to the socio-economic and cultural advancement of the people within the jurisdiction of the University as enjoined by the University Act. He looked upon the Brainstorming Session, organized by the Department of Economics in collaboration with the Institute of Northeast India Studies, as a new beginning in the role of the Assam University in regional development. The concern for development was echoed in the voice of Dr Sumanash Dutta, Head, Department of Economics, Assam University, who proposed a vote of thanks.

In the Presentation Session, chaired by Professor (Ms) Keya Sengupta of the Department of Economics, Assam University, as many as fourteen orations were presented on various facets of the constraints and prospects of development of Barak Valley and the strategies for development in future, viz. 'Continuity of the Colonial Legacy' by Shri Sanjib Deb Laskar, 'Constraints of Development' by Dr Girban Biiswas, 'Post-Colonial Land Economy' by Dr (Mrs) Suparna Roy, 'Exploring Economy for Development Strategy' by Dr Niranjan Roy, 'Economic Development : An Assessment' by Dr Bibhash Ranjan Choudhury, 'Disparities in Economic Development : A District Level Study' by Dr Sumanash Dutta, 'Tea Industry and the Economic Development' by Dr Alok Sen, 'Role of Banking in Development of Agriculture' by Dr (Mrs) Sudipa Rakshit, 'Social Sector Discourse in Development Perspective' by Dr Hironmay Roy, 'Women Participation in Economic Development' by Dr (Mrs) Ruma Paul, 'Border Trade and Growth Rate in Economy' by Dr Ramanuj Chakravorty, 'Exploring Human Resource for Sustainable Development' by Ms Jayati Bhattacharjee, 'Development and Question of Sustainability' by Professor Abhik Gupta and 'Barak Economic Development Council' by Shri Tarun Das. The Chairperson, Professor Keya Sengupta lauded the presenters for the positive approach and observed that the development needs a positive frame of mind which is well reflected in the presentations on various aspects of development. "The problems of development for Barak Valley can be viewed from a multi-dimensional aspect requiring an equally multi-dimensional approach for its solution as revealed by the presentations in this brainstorming session. However, what the solutions will be is ultimately determined by political will, because without political will, economic arguments cannot fall into its place and solutions for development by the academicians may become sterile and meaningless. Along with political will it is also good governance which finally gives shape to the solutions recommended by the academicians and accepted as policies by the state of the political economy in existence." – Professor Sengupta observed.

The Interactive Session turned out to be a serious academic discourse of high order brainstorming under the chairmanship of Professor J B Bhattacharjee in which the participants frankly offered their views on the problems and strategies of development of Barak Valley; raised questions, incited opinions, provoked argumentation, took positions and debated issues with facts and figures, with and without reference to the presentations in the earlier session. The representatives of the 'Barak Upatyaka Banga Sahitya-O-Sanskriti Sanmelan', the 'Barak Upatyaka Bahu-Bhasik Samanway Samiti' and the 'Barak Valley Economic Development Council Demand Committee' claimed away a considerable portion of the time allotted for interaction by strongly arguing in favour of the demand for the institution of an autonomous economic development council for the accelerated economic development of the valley and raised the brainstorming to its logical height by involving the participants in a dialogue on the viability of such an autonomous council. The interactions sometimes generated a lot of heat as the arguments and counter-arguments reached the breaking dots at points of time, and the Chairman intervening with pointed questions and additional information, directing attention to new areas, bringing in alternative proposals and solutions, and inspiring moderation in the tenor of deliberations.

Initiating the interaction, Shri Paritosh Paul Choudhury argued that the main cause of economic backwardness of Barak Valley was the indifference of the governments of the State and at the Centre towards the development of the valley which has been noticed since the partition of the country. The per-capita expenditure in this valley, according to him, was the lowest in the Northeast. He gave example of the two immediate neighbours of the Barak Valley, namely, Mizoram and Tripura. In Tripura, where the total population is equal to that of Barak Valley, the annual budget is of about one thousand crores, and in Mizoram, which has less than half of the population of Barak Valley, it is more than five hundred crores, whereas the Government of Assam hardly spends two hundred crores in three districts in Barak Valley and almost the whole of that amount is expenditure on administration. He felt that low per-capita expenditure was responsible for the low standard of living in the valley. On the other hand, the districts in this valley are important revenue-yielding districts. He referred to a report of 1857 by R Stewart, Superintendent of Cachar (which then included Hailakandi), informing that the revenue generated in Cachar in the previous year was Rupees 1,10,000/-, expenditure was Rupees 60,000-, and the savings of the government was Rupees 50,000/-. The situation has not changed since then. Today also the government collects more revenue from the valley than it spends on administration and development. As a result, the districts in this valley are deprived of the share of the central subsidy to the state for development and welfare activities. The people were apprehensive of this discriminatory attitude on the part of the state government since independence, because it is linguistically and culturally different from the Brahmaputra Valley. Therefore, a memorandum was submitted by responsible citizens to the States Reorganization Commission (SRC) in 1954 for the separation of this area from Assam. The Purbachal Movement of 1950s and the movement for Union Territory in 1970s and 80s are parts of history. The people here had to struggle since independence for the protection of their identity and language. They can never forget the martyrs of the language movements. When Assam was reorganized in 1972 by creating Meghalaya and Mizoram, there were demands for the separation of this valley also. When North-Eastern Council (NEC) was created for the development of the Northeast Region, the Union Territory Demand Committee asked for the representation of Cachar (which then included

Karimganj and Hailakandi) in the Council but it was not conceded on the ground that only States and Union Territories could be members of the NEC. The Government of Assam never took up seriously any proposal for the development of Barak Valley with the Government of India or the Planning Commission or the NEC. Such generosity cannot be expected from them either, since the role of that government in the Sylhet Referendum of 1947 is well known to the people of the valley. Therefore, if development has to take place in Barak Valley it must be empowered with autonomy to generate resources, to receive grants from the central government directly and to look after development with accountability, Shri Paul Choudhury argued.

Shri Tarun Das referred to the memorandum submitted by the Barak Upatyaka Bahu-Bhasik Samanway Samiti to the Prime Minister on 14th August 2006 for creating an Economic Development Council for Barak Valley. Tracing the genesis of the memorandum, Sri Das informed that the Bahu-Bhasik Samanway Samiti was formed in a convention of different linguistic groups in the Barak Valley on 18th August 2005 at Karimganj, which also adopted a resolution demanding the constitution of the said Council. The Barak Upatyaka Banga Sahitya-O-Sanskriti Sanmelan, Manipuri Sahitya Parishad, Nikhil Bisnupriya Manipuri Mahasabha, Barak Hindi Sahitya Samiti, Dimasa Sahitya Sabha, Gurkha Dukkha Nibarak Samiti, and the Karimganj District Tribes Union are the constituents of the Barak Upatyaka Bahu-Bhasik Samanway Samiti which is since then functioning as the Coordination Committee of all linguistic and ethnic groups in the valley. In justification of the demand for an Economic Development Council, Sri Das argued that the history of the Barak Valley since the partition of India is a history of discrimination and deprivation. It is the most neglected and backward part of Assam. The development of this linguistic minority area is possible only by an autonomous council constituted with the representatives of all sections of people in the valley.

The case for an Economic Development Council was strongly and eloquently argued by Shri Kishore Kumar Bhattacharjee who observed that the economic condition of the people of Barak Valley has systematically deteriorated over the years since independence of the country due to negligence of the Centre and the State, and the discriminatory policy of the authorities is responsible for the overall backwardness of the area. The productivity in agriculture and tea – the two mainstays of the economy of the valley – has been going down over the years, but the State Government did nothing to protect the interests of the cultivators and the tea garden labourers and the Central Government was completely indifferent to the economic problems of the valley. While arguing the case of the Economic Development Council, he observed that the North-Eastern Council (NEC) cannot take care of the development of the minority areas within the states for the simple reason that the states are represented in the NEC by the respective governments which are dominated by the majority communities of the states. If the Bodoland Territorial Council (BTC) could be established in the Brahmaputra Valley and the Gurkha Hill Council in Darjeeling, there should be no constitutional or legal difficulty in forming the Economic Development Council for the Barak Valley in Assam, Shri Bhattacharjee opined. He also clarified that the proposed council shall be a non-political body to look only after economic development, although it has to be created through the political processes, and all MPs and MLAs from the valley and the Presidents of the Zilla Parishads in the valley, irrespective of party affiliations, should be the members of the council. Shri Bhattacharjee further informed that the Barak Valley Economic Development Council Demand Committee has already been formed as a federal organization with District Committees in each district in the valley (namely, Cachar, Karimganj and Hailakandi) and a Central Committee in Silchar to pursue the demand for the Economic Development Council.

Dr. Ramanuj Chakravorty felt that it has not been possible to explore the potential of the Barak Valley for sustainable development due to the poor infrastructure, whereas the infrastructure could not be developed because of the discrimination in the allocation of funds received by the state for infrastructure development through five-year plans and under various schemes. The same discrimination is noticed in distribution of funds by the NEC to which the projects are submitted by the State Government. There are also complaints that the proposals for the development of Barak Valley, even if submitted, are not pursued seriously and the funds meant for this valley are diverted to other areas Therefore, Barak Valley needs an autonomous economic development council so that the projects can be directly submitted to the Planning Commission and other authorities and the funds so received can be utilized properly.

Professor Debasish Bhattacharjee introduced a new dimension in the tenor of discussion by suggesting that the Barak Valley deserves to be a state on the ground of economic discrimination and for the protection of cultural identity of the people of the valley. The revenue and the population of the valley would justify the claim, if compared with some of the states within the Northeast. The problem, however, is the contradiction within the local society which was exposed not only during the Sylhet Referendum before Partition but also during various democratic movements since independence. He, therefore, felt it to be important for all the communities within the valley 'to adopt the valley as own land' and to come together and develop a 'we-ness' among the people for the success of the democratic movements for any kind of development. Professor Bhattacharjee also clarified that he was not against the demand for an economic development council, although he felt that the valley deserves to be a state. However, Dr (Ms) Jayati Bhattacharjee felt that the demand for a 'state' at this stage may be difficult due to contradictions within the local society as explained by Professor Debasish Bhattacharjee. In her opinion, the demand for economic development is 'politically neutral' and it concerns all, whereas the statehood is a 'political issue' which might alarm the 'political adversaries'. Moreover, 'unfavourable political culture' and 'contradictions in the local society' may not augur well for a common movement for political autonomy. On the other hand, the demand for economic development is already a common demand of all sections of people in the valley and the benefit of such development will automatically go to all. Therefore, supporting the demand for the economic development council, Dr (Ms) Bhattacharjee observed that as the grass-root participation is needed for development the people in the valley should be involved in the proposed council through the representatives of all the communities. Shri Kishore Kumar Bhattacharjee intervened to add that the success of the economic development council may clear the ground for political autonomy or statehood, but that demand can be considered only when all sections of people in the valley come forward to demand a state. They are now only asking for an economic development council and it is a common demand of the people of the valley. The Barak Valley Economic Development Council Demand Committee has been formed after assessing the public opinion in the valley and the Committee is getting favourable response from the people of the valley in all its programmes, he emphatically claimed.

The discussion turned out to be a serious academic debate when some of the participants made comments, raised questions and sought clarifications about the concept, powers and functions and composition of the proposed economic development council. Dr Alok Sen, for example, observed that comparing per capita expenditure in Barak Valley with Tripura and Mizoram may be out of place because those are states and have their annual budget and accounts and requirement for the cost of governance, whereas such data will neither be available nor applicable in case of the districts. He also pointed out that all the states in the Northeast, including Assam, are special category states, since these states are economically backward as compared to the rest of the country. The Government of India is indeed liberal in providing funds for the development of the region through the North-Eastern Council. The Northeast shares about 8% of the geographical area of the country and about 4% of its population, but each central ministry is now required to spend at least 10% of the budgeted fund on the region. In fact, the Northeast, like each region in the country, needs integrated development. The Barak Valley is a small geographical area within the Northeast and it can be developed only by including the neighbouring areas like North Cachar Hills, Jaintia Hills, Manipur, Mizoram and Tripura within common development schemes, since the resources are spread over the areas and these are complimentary to each other. He also expressed doubts as to whether the MPs and MLAs, who are proposed to be ex-officio members of the economic development council, would be able to rise above the party lines and work for the common benefit of all sections of the people. The allegations of nepotism and favouritism in utilization of the 'local area development fund' of the MPs and MLAs are very common throughout the country, and one of the most important reasons for poor quality of the works is said to be the rampant corruption at various levels. "Unless the character changes there is no hope of development', Dr Sen observed.

Professor Abhik Gupta raised a set of queries to suggest that (i) the inclusion of MPs and MLAs may create the problem of the resolution of conflicting political interests of the parties they represent, (ii) deciding development priorities and fixing the criteria for allocation may cause conflicts between unrepresented micro areas within the valley, (iii) the fair distribution of opportunities among different communities may be another vulnerable area, (iv) defining the 'communities' itself may be a sensitive issue, (v) the power devolution vis-à-vis panchayats may provoke resistance, and (vi) the panchayat experience might call for guard to ensure proper utilization of funds by the economic development council. The message Professor Gupta wanted to convey was indeed loud and clear. "It is necessary to be convinced about 'what is to be achieved' and 'how to be achieved' before hand", he said.

Dr Girban Biswas wanted that the demand for a state or a development council must be a peoples' voice in the valley before it is insisted upon by any organization and that the scholars can help the cause of equitable development in the state by probing into the complaints of discrimination by their research. In his opinion, the greatest weakness for Barak Valley was the absence of a civil society initiative in matters of development. This is a crucial are where the intellectuals of the valley are expected to play a crucial role by sensitizing the people.

Sri Tarun Das, Shri Kishore Kumar Bhattacharjee and Dr (Ms) Jayati Bhattacharjee spontaneously responded to the queries relating to the economic development council made during the course of the discussion. Shri Das was of the opinion that the fact that the resolution demanding the council was adopted in a 'bahubhasik' (multilingual) convention and that a 'samanway samiti' (coordination committee) consisting of the representatives of the different ethnic and linguistic organizations of the valley was formed in the same convention to pursue the demand do amply prove that the demand for the economic development council is the demand of the people of the valley. Elucidating the point, Shri Kishore Kumar Bhattacharjee reiterated that the Barak Valley Economic Development Council Demand Committee is getting favourable response from all sections of people in pursuing the demand. The political leaders from various parties, academicians, youths and students, and representatives of the ethnic, linguistic and literary bodies are involved in the opinion building programmes. The office of the Central Committee is functioning in Silchar and a public memorandum containing history and geography of the Barak Valley, causes of economic backwardness, justification for economic development council, the composition, structure, powers and functions and the sources of revenue of the proposed council, long-standing common demands of the people of the valley, etc, is being published and shall be released on 25th May 2008. The release of the publication shall be followed by a seminar on 'Strategy of Economic Development of Barak Valley' which shall be addressed by important political leaders, eminent academicians and the spokespersons of the social and cultural organizations of different communities in the valley. As regards discrimination, Shri Bhattacharjee said that the people in Barak Valley are deprived of the 'right to information' also, because despite best efforts of the Demand Committee it has not been possible to know the actual amount of revenue collected by the government of the state from the valley. However, according to moderate estimates the revenue contributed by the three districts in the valley should be more than three hundred crores, where as the total expenditure is much less than that amount. "Where is the share of the central subsidy to the state or of the development schemes?" - he asked. He further added that genuinely important development projects for Barak Valley are either not forwarded or included in the five-year plan proposals of the state, and if submitted, these are not seriously followed up. Shri Bhattacharjee also argued that the economic development council shall have to be constituted by a political process, under the provisions of the Constitution of India, taking into account the Pachayati Raj, by the Parliament or the Legislative Assembly or the both, and the legislation or resolution constituting the council should provide for its composition, powers and functions. However, the council should be an autonomous body by way of decentralization within the state, and there are already such development councils for some of the ethnic areas in the Brahmaputra Valley. In his opinion, there is no difficulty in involving the MPs and MLAs, though they may belong to different political parties, because they are the elected representatives of their respective constituencies and they are supposed to work for the development of the constituencies, irrespective of the percentage of votes polled in favour of the elected candidate. Dr (Ms) Jayati Bhattacharjee added that in a democracy with a multi-party system of governance it is but natural that the members of the legislatures shall belong to different political parties; but they do take decisions and make legislations. That is how the Parliament and the Legislative Assemblies do function in this country. The proposed council is also expected to adhere to the same democratic norms.

The Chairman at this stage intervened to rule that there has been a lot of discussion on the suggestion for instituting an economic development council, which may be one of the strategies of development and, if created, the council can be an important instrument of development, but the other strategies should also be explored. Responding to the call of the Chairman, the participants by a voice suggested that the strategies proposed by the presenters in the forenoon session were relevant for the economic development of Barak Valley and all the proposals should receive due consideration. However, Dr. Girban Biswas proposed that the need to control floods should be highlighted as the recurrence of food is a perennial problem and a real menace to the economic development of Barak Valley. The multi-purpose Barak Dam project was envisaged to contribute to agricultural and industrial growth of the Barak Valley and the adjoining areas by converting the flood menace into a source of power generation, but the implementation of the project has been inordinately delayed, first, by indifference of the authority and, then, by provoking political hiccups by the vested interests. "The Barak Dam is a must for the economic development of Barak Valley", Dr Biswas said. Dr S B Paul added that the problems like flood menace and power shortage can be resolved only through modern science and technology. In fact, science is contributing in every sector of development in the contemporary world. Therefore, the promotion of science education and establishment of scientific institutions in the valley should be considered as an important strategy for the economic development of Barak Valley. He also proposed better infrastructure for the existing educational institutions and improvement of the communication network as essential preconditions for development.

The day-long brainstorming session concluded late in the evening with a Valediction, chaired by Professor Tapodhir Bhattacharjee, Vice-Chancellor, Assam University, that excelled in vibrant positive attitude towards the problems of development and the scientific approach and temper in assessing the prospects and proposing the strategies for development. Delivering the Valedictory Address, Professor G D Sharma, Pro-Vice Chancellor, Assam University, proposed the use of contemporary scientific knowledge for the economic development of Barak Valley. The rich biodiversity of the area should be converted into resource for economic development, since there is enough scope for development of agriculture, tissue culture and horticulture in Barak Valley by use of scientific methods and fertilizer. The bamboo related economic activities should be encouraged as this area is rich in natural bamboo forests, and the scientific plantation of bamboo can yield enormous dividends. The solar energy should be harvested, and the 'green urban development' should be an important agenda. All these prospects and potentials of the area should be taken into account in proposing the development strategies. The promotion of science and technology is very important for modern economic development and this requires a scientific temper and a change of the mind-set of the people. Therefore, the science departments in Assam University should be strengthened and these departments should play an important role in the development processes by making the scientific research relevant to the region and by integrating the field studies and extension work into the teaching and research programmes wherever these are feasible. There should also be efforts to establish more scientific institutions and local centres of the national level institutions in the Barak Valley, Professor Sharma observed.

Professor Gopalji Misra, Dean, School of Social Sciences, Assam University, in his address lauded his colleagues in the Department of Economics for the initiative in involving the academia in the economic development of Barak Valley discourse by organizing the brainstorming session. He recalled that the Haryana-Punjab Development Council, which was established at the initiative of Jawarhalal Nehru, then Prime Minister of India, and played the key role in the economic development of that part of the country, in its first meeting decided to develop the roads first and to connect the region with the rest of the country through a road network. He observed that the roads in this part of the country are not very good, although the road between Silchar town and the University Campus has been considerably improved since he saw it first. However, he felt that there is an unfortunate tendency in the area to develop 'puja to puja' roads at low costs so that repair and improvement work becomes due before the next 'puja', whereas the development on long term basis would ultimately be more cost-effective. He also expressed his concern over the poor supply of power in the valley and suggested that the improvement of the power sector should get priority in the development strategies because, besides domestic consumption, power is essential for modern agricultural and industrial growth. Finally, Professor Misra suggested that in Indian situation, besides proposing development strategies, there is always a need for a pressure group to persuade the political authority to approve and implement the development projects. However, it is not necessary that the pressure group should be politically oriented, since the academia can exert enough pressure through research and publications.

Expressing his joy at the high quality of deliberations in the brainstorming, Professor J B Bhattacharjee thanked the authority of the Assam University for their support in organizing the brainstorming session, and the participants for their wholehearted participation. He stated that he always had great regard for the intelligentsia of the Barak Valley, which has now been further reinforced by the presence of the Assam University, and this has been well reflected in the balanced expressions of views in the brainstorming. This reinforced intelligentsia is best positioned to decide what is good for the Barak Valley and what should be the appropriate strategies for its development. He regretted that the valley often misses opportunities by default because as part of Assam its special problems as a landlocked but ethno-linguistically and culturally distinct area within the state as well as its potential of growth as the heartland of the Northeast with a rich human resource reserve do not come into national focus. As per its charter, the Assam University is supposed to play a role in the socio-economic, cultural and intellectual development of the area within its jurisdiction. That the University can do it very well by involving the intellectuals in and outside the University, this brainstorming session is an example. He expected that there shall be several such brainstorming sessions, seminars and workshops on various socially relevant issues in future to highlight the problems and potentials of Barak Valley not only for the localized development of the area but also in regional and national interests. "A Progressive Barak Valley in Progressive Northeast and Progressive India should be our ultimate dream", he said.

Professor Tapodhir Bhattacharjee, in his presidential remarks, reiterated his position in the inaugural session as he said by way of presidential remarks that the Assam University has to play a proactive role in the development processes in the region by maintaining the institution-society linkages and conducting socially relevant research. The development strategy for the area is one such socially relevant research field and, therefore, the University will in future also positively respond to proposals for collaboration with the Institute of Northeast India Studies or any other academic organization in the interest of the people of the area within the jurisdiction of the University. The experience of this brainstorming session has been very encouraging and the concerned departments in the University should come out with similar proposals as did the Department of Economics this time, Professor Bhattacharjee observed. He called the day-long brainstorming a great learning experience in an environment of togetherness of the intelligentsia and a beginning in the process of understanding the roots of economic backwardness and finding solutions.

APPENDICES

Appendix - A

Barak Valley - Basic Statistics 2001

A : District-wise Distribution of Population

District	*Population 2001*			*Growth Rate*	*Sex Ratio*		*Density*	
	Persons	*Males*	*Females*	*1991-2001*	*1991*	*2001*	*1991*	*2001*
Cachar	1,442,141	741,580	700,561	18.66	932	945	321	381
Karim-ganj	1,003,678	516,408	487,270	21.35	946	944	457	555
Haila-kandi	542,978	280,912	262,066	20.92	929:	933	338	408

[*Source* : *Basic Statistics of NER 2002*, NEC, Shillong, 2002].

B : Urban-Rural Population, and Literacy

		Population			*Literates*		
		Person	*Male*	*Female*	*Person*	*Male*	*Female*
Cachar	Total	1,442,141	741,580	700,561	835,888	481,106	354,782
	Rural	1,240,723	478,582	451,549	498,156	284,654	213,502
	Urban	201,418	103,019	98,399	156,401	83,964	72,437
Kari-	Total	1,003,678	516,408	487,270	559,512	317,584	341,928
mganj	Rural	930,131	478,582	451,549	498,156	284,654	213,502
	Urban	73,547	37,826	35,721	61,356	32,930	28,426
Haila-	Total	542,978	280,912	262,066	266,909	157,474	109,435
Kandi	Rural	497,421	257,601	239,820	231,659	138,610	93,049
	Urban	45,557	23,311	22,246	35,250	18,864	16,386

C : Economic Classification of Population

[C = Cachar K = Karimganj H = Hailakandi]

	Total Workers			*Main Workers*		
	Person	*Male*	*Female*	*Person*	*Male*	*Female*
C	464,022	368,990	95,032	363,006	310,324	52,682
K	305,224	247,461	57,763	234,427	208,609	25,818
H	180,983	137,613	43,370	137,962	116,730	21,232

	Marginal Workers			Non-Workers		
	Person	Male	Female	Person	Male	Female
C	101,016	58,667	42,349	978,119	372,590	605,529
K	70,797	38,852	31,945	698,454	268,947	429,507
H	43,021	20,883	22,138	361,995	143,299	218,696

	Cultivators			Agricultural Labourers		
	Person	Male	Female	Person	Male	Female
C	103,985	91,295	12,690	55,779	46,179	9,590
K	73,598	67,822	5,776	46,266	37,450	8,817
H	62,897	52,898	9,919	25,985	19,325	6,660

	Household Industries			Other Workers		
	Person	Male	Female	Person	Male	Female
C	15,565	7,141	8,242	288,703	224,375	64,328
K	14,013	4,728	9,285	171,346	137,461	33,885
H	5,281	1,674	3,607	86,900	63,716	23,184

Appendix - B

Memorandum submitted by the Bahu-Bhasik Samanway Samiti on Economic Development Council

To
The Hon'ble Prime Minister,
Govt. of India,
New Delhi

(Through the Deputy Commissioner, Karimganj)

Memorandum for formation of the Barak Upatyaka Economic Development Council

Respected Sir,

We, on behalf of the Barak Upatyaka Bahu-Bhasik Samanway Samiti, a major forum of the Bengalee and other linguistic communities of the Barak Valley of Southern Assam, beg most humbly and respectfully to submit this Memorandum to your benign self for your kind perusal, judicious advice and necessary action at the earliest.

That Sir, the present Barak Valley consists of the truncated parts of the Kachari Kingdom, annexed to the British India in 1832 as well as the Sylhet district of the undivided Bengal, incorporated with Assam by the British rulers in 1874 for making the newly-formed Province of Assam economically viable. The districts of Cachar and Sylhet were popularly known as the Surma Valley in Assam under the British rule and Assam was made up of two distinct valleys: the Assamese dominated Brahmaputra Valley and the Bengal-populated Surma Valley. The present Barak Valley is the truncated segment of the erstwhile Surma Valley. The Barak Valley had undergone the historic tragedy and trauma of the Partition, when through a referendum the larger part of the Sylhet district went to the erstwhile East Pakistan in 1947 and only three-and-a-half 'thanas' of the erstwhile Karimganj sub-division of the Sylhet district were awarded to India and became associated with the then Cachar district through 'Radcliffe Award'. Since then, the region has to tackle many socio-economic problems, one major problem being the refugee influx from the erstwhile East Pakistan in the early fifties and after.

That Sir, the present Barak Valley with an area of 6,922 sq.km. and a population of 29.89 lakh, as per the census report of 2001, has a higher density of population, which is 448 per sq.km., than that of the rest of Assam, which is 380 per sq.km. The geo-political importance of the Barak Valley is attested by its being an international border with Bangladesh. Though Bengalee forms the major linguistic community in the Barak Valley, other minor linguistic communities, speaking Hindi, Manipuri, Bishnupriya Manipuri, Dimasa, Chorei (Riem Chong), Nepali and Hmars etc. also live here in peace and harmony. But their economic development even after about six decades of independence is almost in a primary stage still.

That Sir, the history of the Barak Valley so far as its economic development is concerned, is a long one. In this context it is worth-while to mention here that V. D. Pandey Committee constituted by the Govt. of India in 1968 earmarked this region as one of the most undeveloped regions of our country. There is no visible change in the sphere of economic growth of the valley since then. Among the major factors, generally held responsible for such lack of development, mention may be made of regular inundation every year, sometimes twice or thrice a year, natural calamities, geo-political situation of the valley surrounded by the international border and the lofty mountains, and very insufficient road, railway and river communications, lack of infrastructure facilities for industrial growth and lack of investment in private sector. The only road communication with the rest of the country runs through the erosion-prone hilly regions of Meghalaya and the meter-gauge rail-communication runs through the hilly region and remains disrupted for days, even months together during rainy seasons. Due to insufficient communication and transportation, the rate of inflation in the valley is at least 20% higher than that of other parts of the country. But the geo-physical and geo-political condition of the valley is not taken into consideration when the national development projects for economic growth of different regions of our country are planned. Hence, the Barak Valley cannot derive any benefit from such projects. Consequently unemployment problem has always been on the rise and there has been a lengthening shadow of gloom and frustration everywhere. All these factors are greatly responsible for the economic backwardness of this region.

The present systems of financing facilities etc are not being easily availed of by the local people because of cumbrous procedure and lack of knowledge.

Considering all the aspects of the situation it is felt that the only way to develop the economic condition of the people of this Valley is to formulate plans for development locally according to the needs of the people and suitability of the region, and it requires formation of a statutory body with specialists as members for chalking out micro-plans for development of this Valley.

That Sir, under the circumstances highlighted above, we beg most humbly and respectfully to urge upon your gracious self to take immediate appropriate measures for setting up an ***Economic Development Council*** for the most neglected

Barak Valley, ensuring the Valley's all-round economic development, for which favour, we shall remain ever grateful and obliged.

Dated, Karimganj (Assam) Yours most faithfully
The 14th August, 2006

[Signed] C. Nilkantha Singha, D. Manglimba Singha & N.Dhananjay Singha on behalf of Manipuri Sahitya Parishad, Assam; Debendra Kr Sinha, Chandra Kanta Sinha & Uday Sinha on behalf of Nikhil Bishnupriya Manipuri Mahasabha; Suresh Chandra Dwibedi, Babul Narayan Kanu & Rabi Nunia on behalf of Barak Hindi Sahitya Samiti; Mukul Ranjan Barman & Biswajyoti Barman on behalf of Dimasa Sahitya Sabha, Edenlal Chorei, Stephen Chorei & Fonidhar Chorei on behalf of Karimganj District Tribes Union; Rajdhan Gurung, Bijay Chetri & Bhakta Singh Thakuri on behalf of Gurkha Dukkha Nibarak Samiti; Dinendra Narayan Biswas, Janmajit Roy & Nitish Bhattacharjee on behalf of Barak Upatyaka Bonga Sahitya-O-Sanskriti Sammelan; and Shyamalendu Chakraborty, President & Tarun Das, Convener, Barak Upatyaka Bahu-Bhasik Samanway Samiti.

Appendix - C

Memorandum of the Barak Valley Economic Development Council Demand Committee

1} Barak Valley Economic Development Council Demand Committee – a brief introduction to its inception and activities

Barak Valley Economic Development Council Demand Committee is an ORGAN of the people, by the people, and for the people of Barak Valley constituted in a Public Convention with thundering voice of all sections of the people of Barak Valley irrespective of communities, castes, tribes, creeds held at Silchar District Library Auditorium on the 18th day of September, 2006. The Central Committee was constituted on 10th June 2007 at Milan Mandir, Sadarghat, Silchar in a delegate session representing all the Administrative Blocks of Barak Valley. In accordance with the provisions of the Committee's Constitution, it has set up the Cachar, Karimganj and Hailakandi District Committees, and presently, we are organizing the Block level/Gaon Panchayat level/ Municipal Ward level committees throughout Barak Valley.

The Committee has been set up against the backdrop of prolonged discrimination, swindle and beguilement by the human parts from Dispur and Delhi, and also from their local agents irrespective of jersey colours and also bureaucrats at apex level in both the State and Centre as a result of which this valley still remains submerged with economic underdevelopment, lack of education, health hazards, acute unemployment, poorest infrastructure facilities, un-utilization/underutilization of natural resources, utterly low GDP and Per Capita Income, lack of proper industrial development specially in agri-field, primitive method of cultivation, humiliating attitude of the nationalized banks and also totally corrupted socio-political devises etc even after half a century of Indian independence. Hence the Committee has to take birth to demand a separate Economic Development Council to finally liberate this valley from the grip of such evil forces.

The Committee has been also, since inception, raising the demands for immediate completion of gauge conversion (Metre Gauge to Broad Gauge) of railway lines from Lumding to Silchar, construction of Dam/Barrage over river Barak and other tributaries in Barak Valley for containment of flood and to provide adequate irrigation facilities as well as to produce electrical energy to meet up the electrical power crisis etc. The Committee has also raised the voice with demands to speed up the work for early completion of the East-West corridor from Silchar to Saurastra (Four Lane Highway), for improvement of Silchar-Guwahati existing two roads, via Jowai and Haflong, for development of infrastructure and communication, for proper irrigation facilities for 3.5 lack hectares of fertile agricultural land, utilization of local resources and to set micro, small, medium and big industry in the valley for the eradication of unemployment and elimination of poverty. The Committee has also demanded for constituting a separate Secretariat of the Assam Government for the valley like other States and also a Circuit Bench of the Guwahati High Court in the Valley.

2) Geography and Eco-Political History of Barak Valley –An Overview

The present Barak Valley consists of three districts, viz. Cachar, Karimganj and Hailakandi. Barak Valley is located in the southern most part of Assam and one of the oldest districts of the State of Assam under the name of Cachar. The District of Cachar (now Barak Valley) is bounded by the hills on all sides, on the north by N C Hills district and Barail

Range, State of Mizoram on the south, State of Manipur on the east, Tripura and Bangladesh on the west. Topographically, Barak Valley is heterogeneous composed of high hills, low lands and plains. Vast tracts in the southern part of the valley are covered with forests. On the north, east and also on the south, the valley is nearly shut in by ranges of hills which in many places occupy considerable parts of the valley's land areas. The two major rivers Barak and Kushiara along with nineteen other tributaries and rivulets flow through the valley.

Historically, a large part of the valley was included in the Cachari Kingdom. That Cachari Kingdom was taken over by the British after the assassination of the last Dimasa (Cachari) King Govinda Chandra Narayan on the 24th day of April, 1830, the last Kingdom to be taken over by the British in India. The District of Cachar (present Barak Valley) was created by the British rulers in 14-08-1832. Throughout the British period, economy of Barak valley was a periphery to the economic mainstream of Bengal. The partition of the country at Independence and the consequent birth of erstwhile East Pakistan and subsequently Bangladesh not only had snapped the traditional cheapest and quickest line of transport and communication, but in the process blocked the age old channels of trade, commerce and transactions. This truncated geography born out of the greatest turmoil in the sub-continent should have attracted special attention of the planners, policy makers and government. But unfortunately, neither the government nor the local political leadership have ever endeavoured to internalize the economic plight at political level. The post independence arrangement did not adequately compensate the valley for all the irreparable damages inflicted on it by the partition leading to a dead loss of political leverage which is otherwise so crucial for economic upliftment of a backward region like this. In 1983 erstwhile Karimganj sub-division and in 1989

Hailakandi sub-division was made separate Districts. The three districts of the State of Assam being situated on the banks of the river Barak and its tributaries together constitute the Barak Valley having total geographical area of 6922 Sq.Km., which lies between Longitude 92.15 and 93.15 East and Latitude 24.8 and 25.8 North. As per the 2001 Census the total population of the Valley was 31,33,966 , and at present the total population is more than 35 lacks as estimated and the density of population of Barak Valley is 340 per sq. km.

3) The Demand for Separate Economic Development Council: Causes Behind

"Necessity is the mother of invention" is the most appropriate words to describe the causes for the demand of a separate Economic Development Council for Barak Valley. An investigative analysis shows that in post independent India, the places geographically remote have remained economically backward as well, until the people living therein succeeded to shake the chairs of the political leaders in power at both provincial and central level with all-out agitation. Under such perspective, it is high time to realize critically that the Barak Valley has also been subjected to utter discrimination in all matters. It has never found any place in the Five Years Planning or in the area development packages so far announced by the Union Government. Electoral politics has played its role in economic discrimination to the valley. The attitude of the Union Government since independence has been of neglect as no comprehensive programme has yet been taken up for economic development of the valley. The attitude of the State Government has been highly discriminatory as observed. It is an undisputed fact that the needs, requirements of the valley selfdom find a place in the planning of Dispur and even if finds, they are few and far between. Even during the devastating floods in 2007, the valley has been given least importance from the point of view of rehabilitation. Thus, when several factors have contributed to the sheer neglect and discrimination as witnessed by the valley since independence AND when the people at large have substantially lost confidence and hope on the present political and administrative set up, and they had to finally, as last resort, fasten their belts to uphold and raise with demands to ensure, implement and enjoy constitutional rights unconditionally through economic powers bestowed at the grass root level by formation of *Economic Development Council* for free, transparent and long awaited development of this valley.

Geographical location of the valley also makes it imperative to raise the demand for special treatment for the valley. Society here is highly cosmopolitan and location is far from the state capital. Even a separate state (Meghalaya) has to be crossed for linking the valley to the state capital. Valley often gets de-linked from the rest of the country. Geographical location actually makes the demand more appropriate.

Lack of proper communication and infrastructure facilities are also contributing to the growing demand for a separate Economic Development Council for this valley. Especially, sheer indifferent attitude and neglect to provide proper communications from and up to this valley and feeling of alienation has fuelled and strengthened this demand over the years. The road and railway communication to and from this valley is pitiable and ramshackle. NH-44 goes through the State of Meghalaya making the travel to the valley a horrible one as well as a costly proposition. Till today broad gauge tracks seem to be day dream for the people of this valley, and there is no hope for the same becoming a reality. People of the valley cannot make use of the railway communication, although railway is said to be the safest and cheapest mode of communication in the country. The available meter gauge up to Lumding often gets encumbered and the available railway communication becomes unavailable more than often. The age old water transport connecting the valley through rivers is blocked for political reasons. The air communication is also not up to the requirements and also very costly in comparison to other parts of the country. All these make the demand of separate Economic Development Council for the valley more viable.

The modern age is age of decentralization. Unilateral decision making and policy implementation have become obsolete in the present age. Decentralization provides easier access to the available opportunities and this is more applicable in case of economic liberation. Decentralization in policy making can make way for formulation of policies in requirement with the peculiarities existing in a particular area and the people living therein. Creation of separate Economic Development Council can ensure more public participation in the development process.

Barak Valley has always been in the fore front in literacy and academic standard. The valley over the years has produced quality individuals. But in the absence of proper employment facilities, it has actually become a favourite place for brain drain. Thousands of educated youths move towards cities and metros from the valley in search of employment every year.

The valley is over populated without means for employment generation. The rise in population has been due to several reasons. As the population is on the rise, resources are on decrease maximizing the number of people below poverty line adding to the complexities and frustrations in the valley. According to the 2001 Census 31,33,966 people inhabiting in the Barak Valley, and at present the total population is more than 35 lacks as estimated; and out of that more than 60% of the people are living Below Poverty

Line (BPL), whereas as per Planning Commission and Central Government, 26% people are living Below Poverty Line (BPL) on the National Level. As per Central Government records the Per Capita Income of North East People including Barak Valley was 4% above the National Level in the year 1951, and in the year 1991 it was 20% below the National Level, and in the year 2006 it was 31% below the National Level.

As per Central Government and State Government records the people of Barak Valley are paying more than 250 (Two Hundred Fifty) Crores per year as direct and indirect taxes to the Central and State Government, but the Central and the State Government are administering this valley by paying far less than the amount so received as direct and indirect taxes. If we study the fund allocation of the Assam Government from 1950 to 2006, we find that the amount allocated for the districts of Barak Valley is far less than in comparison to the other districts of Assam and it reflects a highly discriminatory, parochial and negative attitude of the Assam Government. A comprehensive programme, in conformity with requirements and needs of the people of this valley, can only be a remedy to the rising problems and in the event of establishment of a separate Economic Development Council for the valley empowered to undertake programmes and implement them, it can resolve the crisis situation which the valley is witnessing every day. As the Economic Development Council will consist of representatives of the people, such representatives will be well conversant with the peculiar civic and economic needs of the valley and make programmes count in real sense of the terms.

Agriculture and irrigation of the valley are in an awful condition and facing innumerable difficulties and crisis. Leaving aside the barren tracks, permanent pastures and otherwise cultivable land in the valley are around 40% of the total reported area. It would not be out of place to record the fact here that the national average in this respect is over

60%. Inadequacy of cultivable land on top of a rapidly rising population and unabated agrarian bias of the occupational structure have only meant, progressively, heavier pressure on available agricultural land resulting in excessive fragmentation of holdings and widespread disguised unemployment. The New Agricultural Strategy, popularly known as the Green Revolution of the sixties of this century never made its presence felt in the valley due to lack of initiatives of the government at the centre or in the state. Agriculture here is still on the caprices of tricky nature. As per Assam Government records only 0.2% irrigation facilities are provided to the 3.5 lacks hectares of the net cropped area but practically no irrigation facilities are available in the field. But regretfully, due to non-availability of irrigation facilities coupled with recurrent floods, harvesting has become precariously erratic in this valley. Only in the last August-September, devastating floods, attacking repeatedly as many as four times, left its heavy toll on the valley's economy. With over 15 lacks people of this valley were directly hit by flood and the death toll mounting to 30 (over 50 as per the media estimates) apart from huge loss of livestock and more than 1.5 hectares of crop were destroyed by flood waters. As agriculture here is mostly of mono-cropping variety, there is hardly any chance of recovery before the next rainy season provided no such flood occurs in between. Uneven monsoon has affected greatly and brought down the erstwhile productivity rate of this valley much below both national and state averages. It is pertinent to mention here that up to the nineties, productivity in the field in this valley was well above the national and state averages. The reduction in agricultural productivity is having two fold effects on the people of Barak Valley. The rural people who are engaged in agricultural activities are facing starvation; on the other hand the prices of agricultural products have almost touched skies due to shortage in local production, lack of any sincere effort for supply and proper distribution of such essential items by the respective government heads and their administrative agents holding high positions in this valley. The diminishing curve of agricultural production in the valley is mainly due to lack of government assistance and utter negligence of the government at the centre as well as the state. Here, the need for separate Economic Development Council for the valley gets due leverage.

Since sixty years elapsed from independence, no measures have been taken for industrialization of the valley. All these years, no industry, except Cachar Paper Mill at Panchgram, has been established. The industries existing in the valley from before independence have extinguished gradually due to lack of counseling, technical and financial assistance from the institution of governance. The Sugar Mill at Chargola of Karimganj District is an example of government's total inaction and negligence for industries of this valley. The industrial area of Badarpur Industrial Estate has become a witness to establish how progress of civilization marched backward in this valley due to total indifferent stand of the crooks in power at state and central level. Even in glaring the age of industrialization in state after state, no projects have been undertaken for setting up any industry, irrespective of size, in the valley. Industry is a prime sector of employment. Industries large or small can, apart from economic growth, create employment, both direct and indirect. But industrial development of the valley has been overlooked over the years and still continues to be so, whereas, Barak Valley from all points of view is suitable for industrialization. Peace is a coherent feature of the valley and when the entire state of Assam is hit by insurgency, Barak Valley still remains largely free from such problems.

Law and order situation is much much better than other parts of the state and the country. The valley is rich in raw materials for setting up of at least 300 numbers of Micro, Small, Medium and Big industries. There is no dearth of technically expert professionals in the valley to relate themselves to industrialization and work there in. But inspite of all these advantages, the industrialization of the valley continues to be ignored. Both the centre and the state government have been reluctant, the political forces of the valley have also been largely inactive in raising demands for industrialization.

Since the time British came here, the economy of the valley had largely been dominated by existence of large number of tea gardens in the valley. A large community of workers was brought from the states like Bihar, Orissa, Madhya Pradesh, Uttar Pradesh, Andhra Pradesh and West Bengal to work in those tea gardens. These garden workers have remained in the valley for more than 150 years and thus becoming an inseparable component of population of the valley. The tea garden workers' community is dependent on the tea gardens for their source of employment. There was a time when no scarcity of employment was there in those tea gardens. The gardens owned by British nationals were set up on areas taking on lease for different periods under the Assam Land and Revenue Regulations 1886, which is called the Magnacarta of land laws in Assam. After independence, when British left the ownership the gardens gradually passed on to the business groups having their Head Offices outside the valley and even outside the state of Assam. These new owners of the tea gardens were interested in the accumulation of profits, completely ignoring the research and growth of the tea industry. The accumulated profits were drained out of the valley without any re-investment in modernization of plucking, extension of area of cultivation, installation of modern factory equipment etc. As a result, tea industry in the valley still remains at a primitive stage. But astoundingly, the Government of Assam and the Central

Government have been unscrupulously playing the role of distant observers. No effective steps were taken to stop the drain out of wealth from the valley by means of tea producers. Due to lack of investment from the tea garden management, Govt. indifference etc, the tea industry has been running without investment on modernization making the gardens incompetent to face the competition from the counterparts at national level as well as from other countries, e.g. Chinese and Srilankan. Now the managements of tea gardens are trying to term the tea industries as sick and un-remunerative in the pretext of present crisis in this sector. Some of the smaller tea gardens are actually facing some obstacles in view of the competitive market, but tea production in the valley is still a profitable venture. In the name of crisis, the management of tea producing units in the valley has reduced the number of permanent workers. Some of the gardens are under illegal lockouts, and few have stopped production for management crisis. In simple word, tea gardens in the valley are in mess. Unrest amongst the tea garden workers in the valley can often be witnessed now a days. But till date, the government authorities are not taking any interest to play effective role in resolving the so-called crisis or to adopt welfare activities amongst the workers. A separate Economic Development Council can only identify the core areas for resolving the mess existing in the tea industry and take effective steps to resolve them.

Education of the valley is now in a mess. The valley has a multi-lingual and cosmopolitan population. The methods of education prevalent here are back dated, dominance of Assamese thinking can be observed. The affluent sections have moved their children to the private schools from the government schools. The number of private schools in the urban area is rising day by day; education has become a business in itself. The government schools on the other hand have already lost grounds due to lack of administrative functionaries, widespread corruption in appointments and proximity of fund allocations. The effects are very much apparent in the rural ārcas. Schools in rural areas are in bad shape, teachers are un-attentive to impart modern ideas and values amongst the students of rural areas. The standards have gone down considerably. Education in the valley is becoming more and more difficult for the upcoming boys and girls from the rural areas. Number of drop outs from the primary level is on the increase. The Midday Meal Programme/Sarba Sikshsa Mission has turned out to be a lucrative field of corruption. Instances are plenty where the items of meals, materials and money for the students and schools have actually been exhausted by the authorities-middlemen nexus etc. But neither the Central nor the State Government has the time to give proper look into it or to redress the situation. So, a local authority having the right to interfere and administer the education system in the valley has become quite imperative. A separate Economic Development Council will go a long way in providing the responsive and meaningful education system in the valley.

The basic infrastructure of the valley is in doldrums. The roads are broken, government aided housing facilities are non-existent, slums are on the rise, means of transport and communication are of primitive nature. Crores of rupees is said to have been spent for development of road. But in reality, the roads are in deplorable condition. The condition of National Highway 44 from Silchar to Guwahati is in horrible condition. The other National Highways through the valley too are in pathetic condition. Silchar-Mizoram NH 54 Road, Karimganj-Tripura NH 44 Road, Dhaleshwari-Bhairabi NH154, NH153 etc roads are crumbling. The conditions of other roads of the valley are also in deplorable condition. There are many places in the valley having roads not feasible for running of vehicles and people have to walk to reach urban areas. Such being the condition of roads and communication of the valley, price rise is in high forever and investors are shying away emphasizing the need for a separate set up for promoting economic development of the valley. Another inherent feature of the valley is acute shortage of electricity supply and frequent power cuts. The situation of power sector in the valley is becoming worse day by day. Scarcity of power has dampened many ideas and ventures for setting up small industries in the valley. There are large numbers of villages in the valley where till today people are deprived of electric supply. The requirements are filled up mostly from borrowed sources. The government has never felt any necessity for setting up any power plant in this valley despite scopes. The Assam State Electricity Board often shows discrimination in electricity supply in the valley as compared to other parts of the state. We witness frequent power cuts over long periods of time obstructing the works in offices, small industries, educational institutions, hospitals, shops, residence, centres etc.

A dangerous development over the last few years has been the rise of terrorism and anti-national activities. Lack of employment and resource has severely affected the youth communities resulting in frustration amongst the youths of the valley and they have taken arms in their hands forming terrorist organizations. A responsive and understanding set up is required to understand the difficulties of the youths of the valley and to work for their development. A separate Economic Development Council can actually serve the purpose more effectively than the existing set-ups.

There are examples of setting up separate economic development councils with a view to redress the peculiar issues in such localities. The Bidharva Development Council in Maharashtra, Darjeeling Hills Gorkha Development Council in West Bengal and the Boro Area Territorial Council of our state (Assam) are the prime instances in this regard.

Our Constitution does provide scope for setting up similar councils in the SIXTH SCHEDULE. The Union Parliament has setup such territorial councils by means of amendments of the Sixth Schedule of the Constitution. The Chief Minister of Assam Sri Tarun Gogoi has stated on the floor of the Assam Assembly on the day of placement of Annual Budget 2008-09 that another 6 (six) nos. of Development Councils will be constituted in the State of Assam. The experiments of setting up separate Economic Development Councils have shown positive results in keeping peace and accelerating development of the areas where they have been set up.

4) Demands for Separate Economic Development Council for Barak Valley

Barak Valley Economic Development Council Demand Committee has been set-up with the primary aims and objective of attaining the demand for setting up of a separate Economic Development Council for Barak Valley for achieving the desired goal of economic development of the valley catering to the needs of all sections of people of the valley, and we are strongly placing our DEMANDS for a separate ECONOMIC DEVELOPMENT COUNCIL for BARAK VALLEY on the basis of the reasons mentioned above. The demanded separate Economic Development Council is to be responsible to the people of the valley and fulfill their wishes and desires related to the economic activities. The Economic Development Council is to have full authority to undertake economic development schemes for the valley and to implement the schemes, projects, programmes etc through its organizational setup. On discussion far and wide amongst the members of our organization and in consultation with the leading intellectuals of the valley, we have evolved the following organizational set up, powers and functions, source of revenue etc, for the proposed Economic Development council for the valley.

5) (a) Proposed Organizational Structure of the Demanded Barak Valley Economic Development Council

The Economic Development Council should comprise people's representatives to be elected by public voting in order to represent the thinking, wishes, desires, aspirations of the people of different parts of the valley so that programmes, projects and schemes can actually be for the people. The Council should have a SECRETARIAT with one CHIEF EXECUTIVE OFFICER and also consist of qualified personnel in the fields of engineering, finance, planning, budgeting and legal matters but the general implementation will be made through the existing state government departments, but all the state government employees shall work under the council and their services will be controlled, supervised and monitored by the council.

The State Government will remit the salary, pension etc of the employees to the council after the budget of the each year. The appointment, promotion and transfer etc will be made by the council as per state government rules. The employees working under the council shall not be transferred inside or outside the Barak Valley without the consent of the council.

The Economic Development Council for Barak Valley shall compose of :-

1) GENERAL COUNCIL
2) ONE CHAIRMAN
3) ONE VICE-CHAIRMAN
4) EXECUTIVE COMMITTEE
5) ONE SPEAKER
6) ONE DEPUTY SPEAKER

5) (b) Constitution of the General Council of the Barak Valley Economic Development Council

The General Council shall consist of the members elected directly by means of single transferable votes on the basis of universal adult franchise by the people irrespective of communities, religion, castes, tribes, creeds living in the Barak Valley in specially conducted elections for a term of 5 (five) years. Each of the ZILA PARISHAD Constituency of the Barak Valley under the Assam Panchayat Act, 1987, amended in 1997 shall form constituencies for rural areas, and for town areas some constituencies are to be created in conformity with the number of population of Zila Parishad Area, for election as MEMBERS of the General Council, however the jurisdiction, power and functions of existing Zila Parishad, Anchalik Panchayat and Gaon Panchayat will remain as it is. In addition to these all the MPs, MLAs, Chairmen, Zila Parishads, Chairmen, Town Committees of the Barak Valley shall be ex-officio members of the General Council. The General Council members will elect the Chairman, Vice-Chairman, one Speaker and one Deputy Speaker of the General Council through Secret Ballot.

The General Council shall be in session for 15 (fifteen) days per quarter or as and when required and such session of the General Council shall be called by the Speaker of the General Council. The session shall be called upon when the Parliament and State Legislature are not in session. The members shall debate and discuss on the bills/proposals placed before the house and shall adopt resolutions by members present and voting by a simple majority. For matters found special, the speaker of the house shall form committees and such committees shall place bills/proposals with their recommendations which may or may not be accepted by the members on the floor of the House. The ex-officio members shall also be entitled to take part in the meetings and move proposals and also cast vote, but the ex-officio members cannot hold any post of the Barak Valley Economic Development Council.

5) (c) Election, Power and Functions of Chairman and Vice-Chairman of Barak Valley Economic Development Council

The Chairman and Vice-Chairman of the Barak Valley Economic Development Council shall be elected by the members of the elected General Council through secret ballot. The Chairman is the Chief Administrator and Executor of the Council. He is responsible to lead and guide the council in the proper manner as per the laid down rules, norms etc. The Chairman may nominate members in the General Council from the micro communities living in the valley and unrepresented in the General Council. The Chairman is to prepare the bills, proposals etc and place the same before the General Council. The Chairman will be assisted by the Vice-Chairman in his work and the Chairman will allot the work to the Vice-Chairman and Executive Members. The Chairman may change the allotment of work of the Vice-Chairman and the Executive Members if he finds that the work of the Vice-Chairman and Executive Members are not satisfactory. The Vice-Chairman and Executive Members are to assist the Chairman in his works and work under his guidance and instruction. Chairman, Vice-Chairman and Executive Members shall be responsible to the General Council. The Speaker/Deputy Speaker shall be elected from amongst the elected members. Speaker/Deputy Speaker shall preside over the meetings of the General Council and cast his vote only in case of a tie.

5) (d) Election and Functions of the Executive Committee Members of the Barak Valley Economic Development Council

The Executive Committee of the Economic Development Council shall be the steering wheel of the Economic

Development Council. The number of members of the Executive Committee shall be 15% of the members of the General Council. The Executive Committee shall be headed by one Chairman and one Vice-Chairman who shall be called the Chairman and Vice-Chairman of the Council and be the heads of the Economic Development Council of the Barak Valley. The Chairman, Vice-Chairman and Executive Members shall be elected by the General Council Members on the floor of the General Council by raising hands or through Secret Ballots. The General Council shall have the power to remove the Chairman, Vice-Chairman and Executive Members by placing the no-confidence motion with 2/3 (two-third) members present and voting in the General Council. The Chairman, Vice-Chairman and Executive Committee shall be responsible to the General Council. The Executive Committee shall implement all decisions, resolutions of the General Council and shall be responsible for executive functioning of Economic Development Council and implementation of plans, programmes, projects, schemes, etc undertaken by the council and place reports before the General Council for scrutiny and approval.

5) (e) The Election and Functions of the Speaker/ Deputy Speaker of Barak Valley Economic Development Council

The General Council members shall elect One Speaker and One Deputy Speaker of the General Council through Secret Ballots. The Speaker shall preside over the meetings of the General Council, and shall give his rulings on the Code of Conduct, placing of proposals and adoption of resolutions. The Speaker shall be entitled to debar members from participation temporarily or permanently in the proceedings of the House, if the conduct of any member is found indecent as well as in case of any criminal offence is committed by any member or any punishment is sentenced by any court for any criminal offence.

The Economic Development Council shall have the autonomy to implement all the economic development schemes, plans, programmes undertaken by the Central and the State Government for the valley. The allocated amounts of the Central and State aided Schemes for the valley shall be directly transferred to the Council for disbursement and the Council shall be provided with adequate grants for economic development. Apart from these, the Governments both at the Centre and the State shall provide special budget allocations, plan allocations, grant-in-aid etc. The Council shall be entitled to levy taxes like any other local-self-government. The Council also shall have the right to prepare its plan and programmes according to the requirement and submit it to the Central Government and State Government, Planning Commission and other concerned authorities for approval.

The Council shall be responsible for the economic development, and infrastructure development like road, communication, housing, irrigation, health, micro, small and medium industry as well as agricultural projects and all development works in the valley shall be directly under governance and guidance of the Council. The Council shall forward schemes for infrastructure development of the valley to the Centre and the State Government and implement the approved projects. The Council shall regulate the basic primary and high School level education of the valley. But the Council shall have no role to play in maintenance of Law and Order in the valley. Maintenance of law and order in the valley shall remain with the State Government and the Central Government.

6) Demand Charter of the Barak Valley Economic Development Council Demand Committee

Though Barak Valley Economic Development Council

Demand Committee has been set up primarily to strive and demand for the establishment of a separate Economic Development Council for the Barak Valley, but we can not move away from the other burning realities and demands of the Valley. Hence, the committee has decided to include in its charter of demands of the people which are equally important for the valley, to demand and strive for them. These demands are :-

6) (a) Demand for Immediate Completion of Broad Gauge Railway Line from Lumding to Silchar

The project for construction of the broad gauge railway lines from LUMDING to SILCHAR was sanctioned in the year 1996 Railway Budget with the anticipated cost of Rs 648 crores, but the present anticipated costing is 1676.31 crores, including another 2 Broad Gauge Lines. Due to the delay in execution, the cost is escalating everyday. The works were started more than a decade ago, but till date, there is no sign of progress in completing the work of laying the broad gauge line. As per the N F Railway reports only around 20% of the work was completed up to November, 2007. Innumerable days have been taken, but no tracks are in order. Broad gauge connection is a basic necessity for the valley and not a luxury. But the state government is failing to provide security and the centre is not attaching any importance. Our political leaders are also not vehement in raising their demands in this regard. The Barak Valley Economic Development Council Demand Committee has already written letters to the Union Railway Minister Shri Lalu Prasad Yadav and sought his immediate intervention to speed up the works. It is needed for the people to rise in unison in demand for immediate completion of the project and the Barak Valley Economic Development Council Demand Committee in coming days will make its voice felt in this regard.

6) (b) Demand for Construction of Barak Dam/Barrage over River Barak in Barak Valley

The construction of the Barak Dam/Barrage has been one of the longest standing DEMANDS of the Valley. People of the Valley have struggled for construction of the Dam/ Barrage over river Barak and other tributaries as it is the only hope for the valley to save it from the curse of recurrent floods. Ultimately, the central government has decided to construct a Dam over river Barak at Tipaimukh in Manipur, and the Public Investment Board has cleared the project in the year 2003, and in subsequent meetings dated 25-10-2005 and 31-01-2006. The North East Electric Power Corporation (NEEPCO) has been authorized to construct the Tipaimukh Hydel Project with an estimated cost of Rs 6500 crores under 12th Five Year Plan (2012-2017).

The CCEA authority has not cleared the project as yet as learnt. The original proposal was to construct a Dam/Barrage over river Barak within Barak Valley with the object of containment of flood, generation of electricity and provisions of irrigation facilities. But astonishingly the objective of the dam has been shifted from saving the valley from the curse of recurrent floods and to the production of electricity and provision of irrigation etc. The present location, height etc, has been seriously questioned by the scientists and environmentalists. The present location in fact poses newer threats for the people of the Barak Valley as the valley including the area of the proposed project is in Seismic Zone-5 (five). The Hon'ble Prime Minister of India Dr Manmohan Singh along with other Central Ministers has laid down the foundation stone for Tipaimukh Hydel Project on 16th December 2006 under tight security. The people of Manipur are opposing the said project and 38 public organizations are organizing agitations against the project and called a Manipur Bandh on 16th December, 2006, i.e. on the day of foundation and boycotted the Prime Minister's programme.

The people of Manipur have demolished the Foundation stone as laid by the Prime Minister on 18th December 2006 as reported in the media.

The Barak Valley Economic Development Council [Demand Committee] has already raised its protest against the location, height etc of the present project of the Barak Dam and submitted a Memorandum in this regard to the Honourable Prime Minister of India and other concerned authorities including all Ministers, MPs, and MLAs of the Barak Valley vide letter No. BVEDCDC/MERDM/2007-08, dated 05-10-2007, demanding construction of BARAK DAM/ BARRAGE at Naraindarh or some other convenient place within the Barak Valley and primarily with the object of containment of floods in Barak Valley, provision of irrigation and generation of power to meet up the electricity crisis of Barak Valley. We will march forward with our demand in coming days with the help and cooperation of the people of Barak Valley along with activists and sympathizer of the country.

6) (c) Demands for Containment of Floods in the Barak Valley

The Barak Valley is surrounded by the hills on all the sides, and it is one of the wettest regions of the country having an average annual rainfall of more than 2500 mm. During the 4 (four) months from June to September every year, it receives more than 1300 mm rainfall during these four months, consequently flood occurs almost every year.

Barak Valley is situated in the bank of river Barak and Kushiara along with 19 (Nineteen) Tributaries, and number of Rivulets flow through this valley

The people of Barak Valley are demanding the containment of flood from 1950. But till date no positive action has been taken up either by the Central Government or by the State Government and we, the people, are the net victims every year. In the year 2007, the people of Barak Valley have faced devastating floods in as many as 4 (four) times and 50 people died apart from damage of the livestock and properties. Sufferings of the people are irreparable.

However, it is DEMANDED that an EXPERT COMMITTEE be setup and as per the recommendations of the expert committee, necessary actions be taken at war footings for containment of flood in the Barak Valley without any more delay.

6) (d) Demand for Industrialization in Barak Valley

Industrialization is a major demand of the people of the Valley. Industries – Micro, Small, Medium or Big – can provide employment opportunities and rapid economic progress. A survey conducted by the Central Government as well as by the State Government which shows that the valley has the potential for at least 300 Nos. of Micro, Small, Medium and Big industrial units in total and such potentials must be explored. The Committee has already submitted a memorandum before the Honourable Chief Minister of Assam for industrialization of the Valley. The Committee is duty bound to raise this demand before all forums of the government.

6) (e) Demand for Improvement and Modernization of Agriculture and Irrigation of Barak Valley

The Barak Valley is having around 3.5 lacks hectares of fertile agricultural land. But agriculture of the valley is in deplorable condition. As per 2001 Census the total population of the valley is 31,33,966 and at present the total population is more than 35 lacks as estimated; and out of that, 89% of the people are living in the rural areas of Barak Valley and nearly 53% of the total working force are by and large engaged in agricultural activities. The substantial parts of the agricultural land of the Barak Valley are cultivated by Small and Marginal Farmers and the same constitute nearly 83% of the farmer families. The average size of the farm is only 0.55 hectares. Apart from this most of the cultivable land remains barren from November to April each year due to lack of irrigation facilities.

As per Food Corporation of India (FCI) the valley was a Buffer Stock Area up to 1990 and the FCI authority collected paddy from the villages of the valley and took it to the other parts of the country. But at present the valley is producing only 54% of its own requirements and for the balance of 46% the valley is dependant on other states. As per Assam Government Records, only 0.2% Net Cropped Area is covered under irrigation in Barak Valley but in reality, meagre percentage of irrigation facilities is also not available in the field. There is no scarcity of water in the valley; but due to lack of perspective planning and Govt. assistance there is hardly any development in this basic sector. Governments at the centre as well as in the state will have to be awakened by strong demands and agitations in this regard. We intend to take a leading role in raising the demand for a perspective planning with priority action for 100% irrigation facilities and sanction of adequate fund by the Central and State Govt. and to develop and modernize the agriculture and irrigation sector of the valley without any further delay.

6) (f) Demand for Modernization of Tea Industry of Barak Valley

The Tea Industry of Barak Valley has been an important component and said to be the backbone of the economy of this valley. But now the industry itself is facing innumerable problems, mostly implanted purposefully to achieve vested goal. Hence, there is deep anguish and discontentment amongst the historically tested tea garden workers. The tea cultivation in Barak Valley was started way back in the year

1855, with the indigenous tea sapling without importing the tea sapling from Assam and China. It is the major industry in this valley, around which its economy revolves. At present this industry produces around 49119 thousand kg of tea annually and provides employment to more than 2 (Two) lacks of people directly and indirectly. But, out of that, number of regular workers is only 76,366 as on 31-03-2007. The 5% of the total tea exports of the country is from Barak Valley. More than 100 (One Hundred) tea garden based industry is facing a grim future amongst which as many as 28 gardens are in the sick list and out of which some are already closed and in consequence, it rapidly becomes a threat to demolish the present economic scenario and ecological balance of this valley and also leading to loss of job and retrenchment of a large number of tea workers primarily due to self declared liberty and lordship by the owners of such gardens/industries.

During the 19th century, Cachar (present Barak Valley) Tea was holding the top position in production amongst the districts of Assam. In 1869 Cachar had 24151acres of land under tea cultivation and produced 4,225,000 Lbs. In 1882, the area under cultivation in Cachar had raised to 48,473 acres of land and production also raised to 12,721,000 Lbs. This was the highest production compared with any district of Assam at that time followed by Sibsagar, Lakhimpur and Sylhet. In 1865, the Tea of Cachar (Barak Valley) was highly admired at an exhibition in Paris and the Manager of Lalamukh Tea Estate was rewarded for quality manufacture.

The following information reveals the actual picture of Tea industry of Barak Valley:

Year	*Total Tea Gardens*	*Total Area of Plantation* (in Hectares)	*Total Production (in 1000 kg)*	*Total Regular Workers*
1951	132	30445	19827	51280
1991	121	35314	40605	74450
2007	105	34121	49119	76366

It appears from the above facts and figures that the numbers of tea gardens are reducing, but the areas of tea plantation were raised significantly, and the numbers of regular workers are not increased proportionately. The tea industry is a labour intensive industry and as per a private study, the number of workers are in the required level, but the number of casual labourers are more than double of the regular workers and it is assumed that this has been done only to deprive the casual labourers from all statutory protections/benefits with the motive to earn extra profits.

After an intensive study and analysis in question of speedy regress despite having all the potentials available for further development of tea industry in this valley, the following points/factors are observed :

1) In Barak Valley more than 50% of the tea bushes are of uneconomic age group, but in the Brahmaputra Valley less than 25% of the tea bushes are of uneconomic age group.

2) The cost of production of per kg tea of Barak Valley is less than Brahmaputra Valley and it varies from Rs 10-12 per kg.

3) The wages and benefits paid to the tea workers and employees are far less than Brahmaputra Valley.

4) Most of the tea garden owners of the Barak Valley are taking loan against their gardens and investing the said loan amount in other business outside the valley/state keeping no windows open for investment in this Valley.

5) Artificial crisis are craftily created to make out ways for retrenchment/laid off the workers. But, as inevitable consequence, there have been repeated confrontation in between the management and workers in different tea gardens of Barak Valley and

there is strong apprehension that if such deceitful anti-labour practices are not stopped/set right with added sincerity and endeavour, it may give birth to a blood shedding socio-political unrest for which the unscrupulous garden owners are to be responsible for the consequences.

6) Due to the coercive closures on the plea of sickness of the tea industry, the workers are now at the point of STARVATION and DEATH and these facts are almost regularly finding space in the newspapers which however, as we presume, do not get the due attention of the administration and the leaders.

7) As per the view of the owners of those tea industries, the Auction Prices at the Guwahati Tea Auction Centre are less and they are incurring losses. But this statement is simply a white lie because 29% of the Barak Valley Tea are sold through Guwahati Tea Auction Centre and 20% are sold through Kolkata Tea Auction Centre and 15% are sold directly with high margin of profits as because the market prices are always more than double of the Auction Prices as well as the Auction Prices are much higher than cost of production as learnt from various sources. It is pertinent to mention here that even fixing of low prices at the Guwahati Tea Auction Centre are also questioned by the various quarters for, suspiciously, either having nexus with bulk purchasers like Hindustan Lever, Tata Tea, Eveready Industries and Williamson Maggor or under the circumstances and pressure of the purchasers with the motive to evade the taxes etc as well as to earn extra profits in native markets.

8) According to those owners or their organization, that they are facing tough competition from Chinese and

Srilankan Tea. But practically the problem is not so acute. As per Central Government statement only 17/18 million kg of Chinese and Srilankan Tea are imported per year, and the per capita consumption of tea in India rose from 650 gram to 770 gram and there is a raising trend. Further, more than one hundred crores of people and vast internal market, this small quantity of imports is not giving any effect in the market. Assam is alone producing 52% of the country's gross tea production and INDIA alone produces 30% of the world tea production and more than 23% of the world tea production being consumed by the Indians.

9) The frequent changes of management and ownership as well as division of property among the owners are also a prime factor to cause sickness in tea industries of this Valley.

As per Dr Jayanta Madhav Committee Report some of the small tea gardens are actually facing financial crisis. However, the law and order problem is almost congenial in Barak Valley compared to the Brahmaputra Valley.

Hence, as per the report of the said committee headed Dr Jayanta Mudhav, the Advisor to the Chief Minister of Assam (Economic and Financial) and also in consideration of the above stated factors, it is clear that there is no such serious problem as being focused by the ownership of those tea industries and the problems are more artificial and created as well as propaganda than actual to continue their monopoly and control over the market with the motive to earn more profits and for getting easy passage for diversion of fund elsewhere outside the valley exclusively for their personal gains.

The attitude of the government towards the tea gardens of the valley has been apathetic. Management has been given free hand for exploitation of resources including human resources for draining out profits without any reinvestment for modernization and expansion of the existing industries.

In recent years, the exploitation has reached further heights in the name of CRISIS; number of permanent workers and employees has been reduced considerably. People are engaged for years without being made permanent in their posts to escape the labour regulations.

Considering all the above factors and circumstances, we are placing the following DEMANDS:

a) The Government in the Centre and the State are to take appropriate steps immediately and allocate required fund for modernization of tea industry of Barak Valley.

b) The Central Govt. is to take necessary steps to establish a Tea Auction Centre at Barak Valley like other states.

c) The respective owners of the tea industries are to reinvest 50% of the Annual Profits for modernization/ extension of area of plantation of the tea industry.

d) Since the works of the tea industry are perennial in nature, labourers are to be absorbed as regular employees as per labour laws and regulations.

e) As per records, abundant lands are unutilized within the tea garden area and also outside the garden area under the Govt. as KHAS land. The Govt. and the tea garden owners are to take initiative and allow the unemployed to start the Small Tea Plantation on the unutilized land to become self-employed. The Govt. is to provide infrastructure and required finance to start the Small Tea Gardens in the Barak Valley like the Brahmaputra Valley.

f) Since there is no dearth of educated and technical persons in the tea garden area at present, all the vacancies are to be filled up by the unemployed youth of the tea gardens from this Valley.

g) All the statutory protections and privileges are to be provided to the tea garden workers and employees.

6) (g) Demand for Separate Secretariat, Separate Circuit Bench of Guwahati High Court, and Administrative Tribunal for Barak Valley

The people of this valley, in thousands every day, have to move to Guwahati crossing 300 kms incurring huge expenses for the purpose of the government works and also for filing/defending any legal proceedings before the Guwahati High Court/Administrative Tribunal. Separate Secretariat, separate Circuit Bench of Guwahati High Court and Administrative Tribunal for this valley could save considerable time and energy of the people of this valley. Secretariat in Amritsar/Pune etc and other such establishments in the country are prime examples before us. Justice at the door steps of the people can be achieved if the separate High Court Bench and Administrative Tribunal are established in Barak Valley. As per recent statistics, almost 35% of cases of the Guwahati High Court are from this valley and this simple figure makes it imperative for the valley to have a Circuit Bench here. Lawyers' community of the valley has been vocal in their demands. We believe that public agitations only can achieve this demand. We stand for these demands and will struggle for achieving the same.

6) (h) Demand for Immediate Completion of Silchar-Sourastra Four Lane Road (East-West Corridor)

The Central Govt. has sanctioned the Silchar-Sourastra Four Lane Road (East-West Corridor) in the year 2004 and the work has also started, but the progress of the work is very slow. The State Govt. has failed to provide the land in time and still substantial quantum of land is to be acquired. It is true that the project could not be completed within its scheduled date and if present rate of progress is continued it will not be completed within next five years. The completion of the East-West Corridor is going to reap tremendous benefits not only for people of the valley but it will also give a substantial help for development of the other states adjacent to the Barak Valley, viz. Mizoram, Manipur and Tripura. But the present progress of the works has been astoundingly slow causing inordinate delays. We are raising our DEMAND for immediate completion of the Silchar-Sourastra Four Lane Road (East-West Corridor).

6) (i) Demand for Re-Introduction of Rail, Road and Water Communication up to Kolkata via Bangladesh from Barak Valley

The re-introduction of rail, road and water communication up to Kolkata via Bangladesh from Barak Valley will be revolutionary for the economic development of Barak Valley. The Govt. of India has already taken such steps for other states of the country. We are DEMANDING that the Govt. of India and the state of Assam are to take positive steps in this regard for the interest of Economic Development of Barak Valley and the state of Assam.

6) (j) Demand for Modernization of the Silchar Medical College and Hospital

The Silchar Medical College and Hospital was established along with other two Medical Colleges of the state. But when the other two Medical Colleges of the state have been enriched with all modern infrastructures, equipments, doctors, facilities etc, the Silchar Medical College and Hospital still remains far behind them. Discriminatory and the step motherly behaviour of the State Government are responsible for the present position of the Silchar Medical College and Hospital.

The Silchar Medical College and Hospital has been catering to needs of the people of not only Barak Valley but also to the people of the adjacent states like Tripura, Mizoram and Manipur. Thousands of patients move into the S.M.C.H. everyday for treatment as indoor and outdoor patients. But the facilities here are not up to the mark or in conformity with the modernization that has taken place in the health sector. S.M.C.H. is lacking in modern equipments, doctors and technicians and also other basic facilities. Immediate modernization of the S.M.C.H. by improving the basic facilities, installation of modern equipments and employment of adequate number of doctors and qualified technicians are of high importance and we DEMAND the same.

6) (k) Demand for Immediate Improvement of Existing Silchar-Guwahati Roads

There is a dilapidated road in the name of NH-44 connecting Silchar with Guwahati via Meghalaya (Jowai, Shillong) through the high hills, which is the life line of the people of Barak Valley as well as the people of Mizoram, Tripura and a part of Manipur. But this road frequently remains closed for days together during the monsoon time i.e. for around 6 months in a year. Apart from this, it is also very dangerous and risky road through the high hills and accidents occur regularly causing loss of lives and properties.

There is already another alternate road connecting Silchar with Guwahati, comparatively shorter in length, running through the State via Udarbond, Harangjao, Turuk, Panimore, Nelly, Jagiroad. But it is not useable due to lack of improvement of a small portion of the road despite the fact that this road is shorter, safer and cheaper. Since it is passing through the state of Assam, it will also boost the trade and commerce of the state and thereby enhance its revenue.

The demand for starting of the said road was raised by the various organizations of the valley and the same was also placed in the floor of the Parliament; but there is no positive move in this regard as yet due to sheer indifferent attitude and negligence of the State Govt. We are also raising our DEMAND for immediate improvement of the road in question without any more delay.

6) (l) Demand for setting up of a Hydro/Thermal Power Project in the Barak Valley

The Barak Valley is facing acute power crisis throughout the year since last two decades and the crisis is day by day increasing. The power is supplied to this valley by the Assam State Electricity Board (ASEB). The present requirement of the Barak Valley is around 80 MW, but the ASEB authority (controlled from Guwahati) is not supplying more than 35-40 MW per day. Consequently, there are power cuts for long hours every day and the small industrial units, tea industry, hospitals, educational institutions, and residences etc are the worst sufferers. The crisis of electricity is a major obstacle for growth of new industry in this valley. The supply of electricity by the ASEB authority is inadequate and discriminatory and the various organizations of Barak Valley are raising protest against the ASEB authority regarding discrimination from time to time. The ASEB authority has not taken any steps for development as well as improvement of the power in Barak Valley since inception and the ASEB authority is just pocketing the profit as a colonial ruler.

In view of the present position, we are DEMANDING immediate improvement of the existing system and supply of required quantity of electrical energy till setting up of a HYDRO/THERMAL POWER PROJECT in BARAK VALLEY, as because Barak Valley has the full potential of such type of project, and the people of Barak Valley is to get electrical power on first priority, and in cheaper rate like the tripartite agreement signed in between the Central Government, Manipur Government and NEEPCO authority regarding the proposed Tipaimukh Hydel Project.

6) (m) Demand for setting up of Ancillary Industries based on Hindustan Paper Corporation Ltd, Cachar Paper Mill, Panchgram, Dist. Hailakandi and Demand for Modernization of the Mill as well as for an independent Enquiry regarding alleged Corruption and Malpractices in the Mill

The Govt. of India, Ministry of Heavy Industries and Public Enterprise had initially sanctioned Rs 426 crores for construction of a 300 (Three Hundred) tons per day production capacity paper mill and the Hindustan Paper Corporation Ltd (A Central Public Sector Undertaking) constructed the Cachar Paper Mill at Panchgram, Dist. Hailakandi. The said Paper Mill started its commercial production in the year 1987. The performance of the mill is not at all satisfactory. The mill is suffering from various fronts and every month getting headlines in the newspapers regarding the shortage of inputs, corruption, malpractices etc, and it can be presumed that if the present situation continues, the future of the mill will be bleak.

The Central Bureau of Investigation (CBI) has made a few enquiries regarding the various allegations but they were managed/disarmed, as alleged, from the various quarters. We DEMAND that an independent high power ENQUIRY be made to find out the cause and the culprits as well as action be taken against them.

The Central Government of India, Ministry of Heavy Industries, has advised the Hindustan Paper Corporation Ltd, from time to time to promote the ancillary industries, outputs of which are required by the HPC everyday. As per Central and State Govt. records 16 (Sixteen) Nos. of ancillary industries could be established based on HPC, Cachar Paper Mill. But none of the ancillary industries has come up since last 20 years of starting the commercial production. Hindustan Paper Corporation ((H.P.C.) Cachar Paper Mill provides the scope for many ancillary industries within the valley. The ways and means of establishing such industrial units within the valley must be explored so as to generate economic growth and employment opportunities within the valley.

The technology of the HPC, Cachar Paper Mill is an old one and it urgently requires modernization to reduce the cost as well as to compete in the most competitive market for its survival. It was learnt from the media that the Govt. of India has sanctioned Rs 250 crores in the year 2005 for modernization; but even after passing of 2 (two) years, no positive and effective action has been taken for modernization as learnt. But substantial amount of sanctioned money was already spent improperly in the name of modernization which is alleged from the various quarters and it is apprehended that in the name of modernization the allotted amount may be evaporated.

6) (n) Demand for Modernization of National Institute of Technology (former Regional Engineering College), Silchar, Cachar, Assam

The Government of India had sanctioned a Regional Engineering College at Silchar, in the year 1976 and it is the 14th Regional Engineering College of the country .Since inception the college has been suffering from various difficulties. When other Regional Engineering Colleges of the country have already reached the desired level, the R.E.C. Silchar failed to do so till date, save, up gradation to the level of N.I.T.

Further, when the present engineering education is highly sophisticated everywhere in India, the Regional Engineering College at Silchar is suffering immensely as on date due to lack of modern laboratory, library, infrastructure, etc.

Though the name of the institute is **National Institute of Technology**, the college is running with a few streams of engineering courses when there are so many branches of engineering education being imparted in the country in other colleges. So, we strongly DEMAND that all the modern branches of engineering education like Biotechnology, Chemical Engineering, Tea Technology, integrated courses leading to Ph D Degree in pure/applied sciences, Centre of Excellence in Earthquake Engineering, etc be introduced with allocation of adequate fund for development of infrastructure, required number of laboratories, modernization of existing laboratories, class rooms, hostel facilities, library and provision of I T facilities with standard software etc useful to students.

National Institute of Technology, Silchar is providing quality technical education to the students of North Eastern States and to cater to the demands of the N E Students intake capacity of the existing branches of engineering is to be increased.

A Science and Technology Park may be established to enhance interest among the students (both juniors and seniors and of different standards/levels) towards science and technology.

6) (o) Modernization of Schools and Colleges of Barak Valley

The schools and colleges of the valley are facing innumerable difficulties. The proportion of teachers vis-à-vis students is all time low and fund allocations are scarce. Facilities of library and modern education techniques are also required to be improved. Again, there are hundreds of schools (from L.P. to H.S. level) which are not even provided with minimum basic requirements. So, we raise our DEMANDS to set right the position without any delay.

A good number of Govt. recognized LP, ME, Madrasa, Tol, High School, Junior and Degree Colleges were established in the Barak Valley and imparting education to the students since last 15-20 years without any salary/ remuneration. The teachers of the said schools and colleges are demanding the provincialization of their services since long time. Since the teachers are working without any salary/ remuneration and doing a yeoman's service to the society, we are DEMANDING that their services are to be provincialized without any more delay.

6) (p) Demand for establishment of Agro-based Industry

Barak Valley has potential for growth and development of agro-based industries, provided the institutional assistance in training and finance are given to the youth of the valley. Agro-based industries can become a lucrative alternative source for economic growth as well as for employment generation. We DEMAND for initiation of positive steps immediately towards the growth and development of agro-based industry in this valley.

6) (q) Demand for establishment of Food Processing Industry in Barak Valley

There are ample opportunities for growth and development of food processing industry in Barak Valley in small scale and such industries can be established with minimum investment. Hence, substantial number of these industries can be set up by the Govt. providing necessary guidance, training and support through financial institutions to meet the need for machinery and equipments, working capital, infrastructures etc. We are DEMANDING immediate steps by the Central and State Govt. for establishment of Food Processing Industry in Barak Valley.

6) (r) Demand for Development of Fishery in Barak Valley

The Barak Valley has plenty of opportunities for growth

and development of Fishery. But we can not explore these opportunities due to lack of Govt. assistance. The requirement of fish is very high in Barak Valley compared to other areas and the local production is too little to meet the daily requirements. So, we are to import fish from Andhra, Kanpur and other areas of the country and the cost is around 25 crores per year with increasing trend. We are DEMANDING that the Govt. is to chalk out the plan and programmes for the growth and development of FISHERY in the Barak Valley.

6) (s) Demand for Development of Tourism in the Barak Valley

The tourism has become a money-spinning and rapidly growing industry in modern times. There are ample scope for growth and development of Tourism Industry in Barak Valley. The Central and State Govt. have been taking many steps for development of tourism industry in the state. But our valley is kept behind to get the required importance. We urge upon the Government to initiate appropriate steps for development of tourism in the valley as an industry.

6) (t) Demand for Development of International Trade Centre at Sutarkandi

The Government of India has sanctioned an International Trade Centre at Sutarkandi (near Bangladesh border) in the district of Karimganj of Barak Valley and the work of the centre was started a few years back. But the centre is not developed to the required level as yet. We are DEMANDING that International Trade Centre at Sutarkandi is to be developed in the required level without any more delay.

6) (u) Demand for Natural Gas-based Power Project and Full Fledged Permanent Establishment of ONGC

The Barak Valley has the large store of Oil and Natural Gas as stated by the various Committees, experts and the

ONGC is exploring the crude oil and the natural gas since last few years. But their works are suffering due to lack of modern equipments, machineries etc and their performance is much below the expected level due to utter casual approach.

Natural gas based power projects can provide solutions to the problem of power deficiency in the valley. Since the Barak Valley has the large store of Natural Gas and the same must be utilized for generation of power. We are DEMANDING that the necessary steps are to be taken for setting up of Natural Gas based power project and full fledged permanent establishment of ONGC at Barak Valley.

6) (v) Demand for Disaster Management System

The Central Govt. has taken initiative for disaster management and every district administration has been given instruction in this regard. The Barak Valley is situated in Zone-5 which is highly sensitive earthquake zone. We request the Govt. for separate plan allocations, and comprehensive programme for disaster management which is a necessity for the valley and we raise our voice for the same.

6) (w) Demand for Development and Preservation of Medicinal Plants

There are abundant quantities of Medicinal Plants in the forest areas of Barak Valley and a number of Pharmaceutical Companies are collecting the same in bulk quantity every year. The medicinal plants are of rare variety with high medicinal value. We are demanding for development and preservation of Medicinal Plantation in Barak Valley as well as to preserve the rare variety.

We must reiterate from our history that we never retreated from fight and struggle to earn our objectives, not only for the independence of our country, but also after the independence, we have shed our bloods number of times and some of our brothers and sisters became MARTYRS for the cause of the society.

We are sure that we will get whole hearted support and cooperation from each and every people living in Barak Valley and out side. If the Central and State Government is not giving due importance to our DEMANDS, we will start and continue the DEMOCRATIC AGITATION till the achievement of our goal.

WE ARE JUST ON OUR MARKS FOR A LONG MARCH, WE HAVE GOT ONLY TWO CHOICES – GET OR DIE ECONOMIC DEVELOPMENT COUNCIL FOR BARAK VALLEY

With mass public support and cooperation we will definitely achieve the goals set forth and DEMANDS raised for.

Dated, Silchar
25 May, 2008

(Kishore Kumar Battacharjee)
President, Central Committee
Barak Valley Economic Development
Council Demand Commitee

Appendix - D

Participants of the Brainstorming Session

1. Professor J B Bhattacharjee, Chairman, Institute of Northeast India Studies, 'Barak', 13/A Udayan Abasan, Udayrajpur, Madhyamgram, Kolkata – 700129
2. Professor Tapodhir Bhattacharjee, Vice-Chancellor, Assam University, Silchar – 788011
3. Professor G D Sharma, Pro-Vice Chancellor, Assam University, Silchar – 788011
4. Professor Gopalji Misra, Dean, School of Social Sciences, Assam University, Silchar - 788011
5. Professor (Ms) Keya Sengupta, Department of Economics, Assam University, Silchar – 788011
6. Dr Sumanash Dutta, Head, Department of Economics, Assam University, Silchar – 788011
7. Dr Niranjan Roy, Reader in Economics, Assam University, Silchar – 788011
8. Dr Projit K Palit, Reader in History, Assam University, Silchar – 788011
9. Dr (Ms) Sudeshna Purkayastha, Reader in History, Assam University, Silchar – 788011
10. Professor Ramendu Bhattacharjee, Department of Physics, Assam University, Silchar – 788011
11. Dr Girban Biswas, Retd. Lecturer in Political Science, G C College, Silchar – 788004
12. Dr (Mrs) Susmita Gupta, Sr. Lecturer in Ecology & Environmental Science, Assam University, Silchar – 788011

13. Dr Ramanuj Chakravorty, Sr Lecturer in Commerce, Karimganj College, Karimganj – 788710
14. Dr Hiranmoy Roy, Head, Department of Economics, Karimganj College, Karimganj – 788710
15. Dr Bibhash Ranjan Choudhury, Director, College Development Council, Assam University, Silchar – 788011
16. Dr (Ms) Jayati Bhattacharjee, Reader in Political Science, Assam University, Silchar – 788011
17. Shri Sandip Dey, Public Relations Assistant, Assam University, Silchar – 788011
18. Shri Uttam Kumar Saha, Correspondent, *Dainik Samayik Prasanga*, Silchar – 788001
19. Ms Sarbani Bhattacharjee, Research Scholar, Department of History, North-Eastern Hill University, Shillong – 793022
20. Dr Subhrabaran Das, Lecturer in Economics, Assam University, Silchar – 788011
21. Shri A K Roy, Officer on Special Duty, Vice-Chancellor's Secretariat, Assam University, Silchar – 788011
22. Dr (Mrs) Subhra Nag, Reader in Philosophy, Assam University, Silchar – 788011
23. Dr Ram C Majhi, Reader in Philosophy, Assam University, Silchar – 788011
24. Professor Debasish Bhattacharjee, Department of Political Science, Assam University, Silchar – 788011
25. Professor Haripada Chakraborty, Department of Sanskrit, Assam University, Silchar – 788011
26. Shri Tarun Bikash Sukai, Lecturer in Social Work, Assam University, Silchar – 788011

27. Professor Rama Bhattacharyya, Department of Bengali, Assam University, Silchar – 788011
28. Shri Niharendu Dhar, Section Officer, Assam University, Silchar – 788011
29. Shri Haji Md Abdul Jalil, Section Officer, Assam University, Silchar – 788011
30. Shri S B Das, Assistant Registrar, Assam University, Silchar – 788011
31. Shri Paritosh Paul Choudhury, Journalist, Silchar – 788001
32. Dr S Sengupta, Registrar, Assam University, Silchar – 788011
33. Professor Abhik Gupta, Department of Ecology & Environmental Science, Assam University, Silchar – 788011
34. Dr Rashni Ranjan Mishra, Reader in Law, Assam University, Silchar – 788011
35. Professor Biman Kumar Dutta, Department of Ecology & Environmental Science, Assam University, Silchar – 788011
36. Dr Parthankar Choudhury, Reader in Ecology & Environmental Science, Assam University, Silchar – 788011
37. Dr Subhabrata Dutta, Lecturer in Social Work, Assam University, Silchar - 788011
38. Shri Ritwik Mazumdar, Lecturer in Economics, Assam University, Silchar – 788011
39. Shri Kishore Kumar Bhattacharjee, General Secretary, I.N.T.U.C., Assam Branch, and President, Central Committee, Barak Valley Economic Development

Council Demand Committee, Norshing Tola, Silchar – 788001

40. Dr (Mrs) Sudipa Rakshit, Sr. Lecturer in Economics, Cachar College, Silchar – 788001

41. Dr (Mrs) Ruma Paul, Sr. Lecturer in Economics, G C College, Silchar - 788004

42. Shri Tarun Das, General Secretary, Barak Upatyaka Banga Sahitya-O-Sanslriti Sammelan, and Convener, Barak Upatyaka Bahu-Bhasik Samanway Samiti, Banamali Road, Karimganj - 788710

43. Shri Bishwanath Majumdar, Office Secretary, Barak Upatyaka Banga Sahitya-O-Sanskriti Sammelan, Banamali Road, Karimganj – 788710

44. Ms Pankaj Moro, Lecturer in Economics, Women's College, Silchar – 788001

45. Ms Phalguni De, Head, Department of Economics, Women's College, Silchar – 788001

46. Ms Jayanti Dutta, Lecturer in Economics, Women's College, Silchar – 788001

47. Shri Ratan Deb, Correspondent, *Dainik Yugosankha*, Silchar – 788007

48. Shri Sanjib Deb Laskar, Editor, *Sambhar*, Ullaskar Dutta Sarani, Silchar – 788001

49. Professor G P Pandey, Department of Mass Communication, Assam University, Silchar – 788011

50. Dr S B Paul, Reader in Chemistry, Assam University, Silchar – 788011

51. Dr Alok Sen, Reader in Economics, Assam University, Silchar - 788011

52. Professor Subir Kar, Department of Bengali, Assam University, Silchar – 788011

53. Shri Raju Mandal, Lecturer in Economics, Assam University, Silchar – 788011

54. Professor Tanmoy Bhattacharjee, Department of Political Science, Assam University, Silchar 788011

55. Shri Debojit Dutta, Research Scholar, Department of Economics, Assam University, Silchar – 788011

56. Ms Suchitra Das, Research Scholar, Department of Economics, Assam University, Silchar – 788011

57. Ms Tanaya Paul, Research Scholar, Department of Economics, Assam University, Silchar – 788011

58. Ms Kamauki Konwar, Research Scholar, Department of Economics, Assam University, Silchar – 788011

59. Ms Karabi Gogoi, Research Scholar, Department of History, Assam University, Silchar – 788011

60. Ms Deepali Bhattacharjee, Research Scholar, Department of Economics, Assam University, Silchar – 788011

61. Shri Barindra Das, Research Scholar, Department of Economics, Assam University, Silchar – 788011

62. Ms Paramita Kar, Research Scholar, Department of Economics, Assam University, Silchar – 788011

63. Ms Swagata Das, Research Scholar, Department of Economics, Assam University, Silchar – 788011

64. Ms Anamika Bhattacharya, Research Scholar, Department of Economics, Assam University, Silchar – 788011

65. Ms Sumitra Bithi Kachari, Research Scholar, Department of Economics, Assam University, Silchar – 788011

66. Ms Nabanita Debnath, Research Scholar, Department of Economics, Assam University, Silchar – 788011

67. Ms Jitu Saikia, Research Scholar, Department of Economics, Assam University, Silchar – 788011

68. Ms Paramita Dasgupta, Research Scholar, Department of Economics, Assam University, Silchar – 788011

69. Dr (Mrs) Suparna Roy, Lecturer in History, Rabindra Sadan Girls' College, Karimganj – 788710

70. Ms Jayati Bhattacharjee, Lecturer in History, Women's College, Shillong - 793003

Index